Halima HAJJI
Mohammed Aziz AJANA
Mohammed BOUACHRINE

Combining Medicinal Plants and AI to Combat Respiratory Infections

Halima HAJJI
Mohammed Aziz AJANA
Mohammed BOUACHRINE

Combining Medicinal Plants and AI to Combat Respiratory Infections

Noor Publishing

Imprint
Any brand names and product names mentioned in this book are subject to trademark, brand or patent protection and are trademarks or registered trademarks of their respective holders. The use of brand names, product names, common names, trade names, product descriptions etc. even without a particular marking in this work is in no way to be construed to mean that such names may be regarded as unrestricted in respect of trademark and brand protection legislation and could thus be used by anyone.

Cover image: www.ingimage.com

Publisher:
Noor Publishing
is a trademark of
Dodo Books Indian Ocean Ltd. and OmniScriptum S.R.L publishing group

120 High Road, East Finchley, London, N2 9ED, United Kingdom
Str. Armeneasca 28/1, office 1, Chisinau MD-2012, Republic of Moldova, Europe
Printed at: see last page
ISBN: 978-620-7-47924-5

« Combining Medicinal Plants and Artificial Intelligence to Combat Respiratory Infections »

Authors

Dr. Halima HAJJI

Prof. Dr. Mohammed Aziz AJANA

Prof. Dr. Mohammed BOUACHRINE

"Scientific knowledge is the remedy for

the ignorance that threatens health."

To my Family, Professor and Friends $---$

Thank you for your love and support

Preface

The year 2019 marked a pivotal moment in modern history with the emergence of the global COVID-19 pandemic, caused by the SARS-CoV-2 virus. Since then, the worldwide scientific community has mobilized to understand this virus, its mechanisms of action, and its effects on human health. Among the many avenues explored, medicinal plants have garnered significant attention due to their proven therapeutic properties and historical use in treating respiratory diseases.

This book, « **Combining Medicinal Plants and Artificial Intelligence to Combat Respiratory Infections** », aims to explore the intersections of phytotherapy, artificial intelligence (AI), and computational chemistry in the fight against respiratory illnesses. By examining plants such as **Saussurea costus**, known for its medicinal properties, we seek to discover how these natural remedies can complement modern approaches to treating COVID-19 and asthma.

Saussurea costus, also known as "Costus," has been used since antiquity for its therapeutic virtues. Originating from India, this plant has been widely adopted in various cultures to treat a range of respiratory ailments. During the pandemic, an increase in the consumption of Costus in Indonesia was observed, which coincided with a lower hospitalization rate among Indonesian patients compared to other countries. This phenomenon sparked particular interest in studying the active components of this plant and their potential to combat SARS-CoV-2.

At the same time, chronic respiratory diseases such as asthma continue to pose significant public health challenges. Asthma, characterized by inflammation of the airways, leads to severe symptoms that can be exacerbated by viral infections like COVID-19. In this context, polyphenols extracted from Chinese plants, known for their anti-inflammatory properties, offer a glimmer of hope.

By combining traditional phytotherapy approaches with modern tools of artificial intelligence and computational chemistry, this book explores innovative methods to identify and optimize treatments. AI enables the modeling of molecular interactions and the prediction of the effects of active compounds, thereby accelerating the drug discovery process. Furthermore, computational chemistry helps us understand the mechanisms of action of medicinal plants at a deeper level and ensures the safety and efficacy of treatments.

Our goal is to provide an integrated, multidisciplinary perspective on how medicinal plants, supported by cutting-edge technologies, can contribute to the fight against respiratory diseases. We hope this book will inspire new research and collaborations and pave the way for more effective and accessible treatments for those suffering from these conditions.

We invite you to explore the following chapters, where we detail our findings and analyses and share our vision for a future where modern science and ancient wisdom unite to improve human health.

With gratitude,

Dr HAJJI Halima

Contents

<u>Chapter 1:</u>

Computational approach investigation bioactive molecules from Saussurea Costus plant as SARS-CoV-2 main protease inhibitors using reverse docking, molecular dynamics simulation, and pharmacokinetic ADMET parameters

Chapter 2

Assessment of asthma treatment against SARS CoV-2 by using a computer approach

<u>Chapter 2 :</u>

Assessment of asthma treatment against SARS CoV-2 by using a computer approach

Abbreviations list

QSAR: Quantitative Structure-Activity Relationship

MD: Molecular Dynamic

WHO: World Health Organization

RD: Reverse Docking

PDB : Protein Data Bank

PBC: Periodic Boundary Conditions

PME: Particle Mesh Ewald

TCM: Traditional Chinese Medicine

MW: Molecular Weight

HBD: Hydrogen Bond Donor

HBA: Hydrogen Bond Acceptor

Aeg : Arginine

Asn : Asparagine

Asp : Aspartic acid

Cys : Cysteine

Glu : Glutamic acid

Gln : Glutamine

Gly : Glycine

His : Histidine

Ile : Isoleucine

Leu : Leucine

Lys : Lysine

Met : Methionine

Phe : Phenylalanine

Pro : Proline

Ser : Serine

Thr : Threonine

Try : Tryptophan

Tyr : Tyrosine

Val : Valine

General Introduction

In recent years, respiratory diseases have become a significant global health concern, underscored by the COVID-19 pandemic that began in 2019. COVID-19, caused by the SARS-CoV-2 virus, has led to widespread illness and death, primarily affecting the respiratory system. The virus enters the body through mucous membranes in the nose, mouth, and eyes, leading to symptoms such as fever, cough, shortness of breath, and severe respiratory complications, particularly in vulnerable populations like the elderly and smokers.

The urgency of finding effective treatments for COVID-19 and other respiratory diseases has spurred interest in both natural remedies and modern scientific approaches. Traditional medicinal plants, such as Saussurea costus, have been used for centuries to treat respiratory ailments. This plant, known for its anti-inflammatory and antimicrobial properties, has shown potential in alleviating symptoms of respiratory infections, including COVID-19. The increased use of Saussurea costus during the COVID-19 pandemic in Indonesia, which reported lower hospital admission rates, highlights its possible therapeutic benefits.

Parallel to the exploration of natural remedies, the field of computational chemistry offers powerful tools for drug discovery and development. Techniques such as molecular docking, molecular dynamics simulations, and ADMET (absorption, distribution, metabolism, excretion, and toxicity) studies enable researchers to predict the efficacy and safety of potential treatments. These computational methods facilitate the identification of active compounds in medicinal plants and their interactions with viral proteins, providing a scientific basis for their use in treating respiratory diseases.

This book aims to bridge the gap between traditional medicinal knowledge and modern computational chemistry. By investigating the active ingredients of Saussurea costus and other Chinese medicinal plants through computational approaches, we seek to uncover new potential treatments for COVID-19 and asthma. Through the integration of natural remedies and advanced scientific techniques, this work aspires to contribute to the global effort in combating respiratory diseases and improving public health.

Chapter 1

Computational approach investigation bioactive molecules from Saussurea Costus plant as SARS-CoV-2 main protease inhibitors using reverse docking, molecular dynamics simulation, and pharmacokinetic ADMET parameters

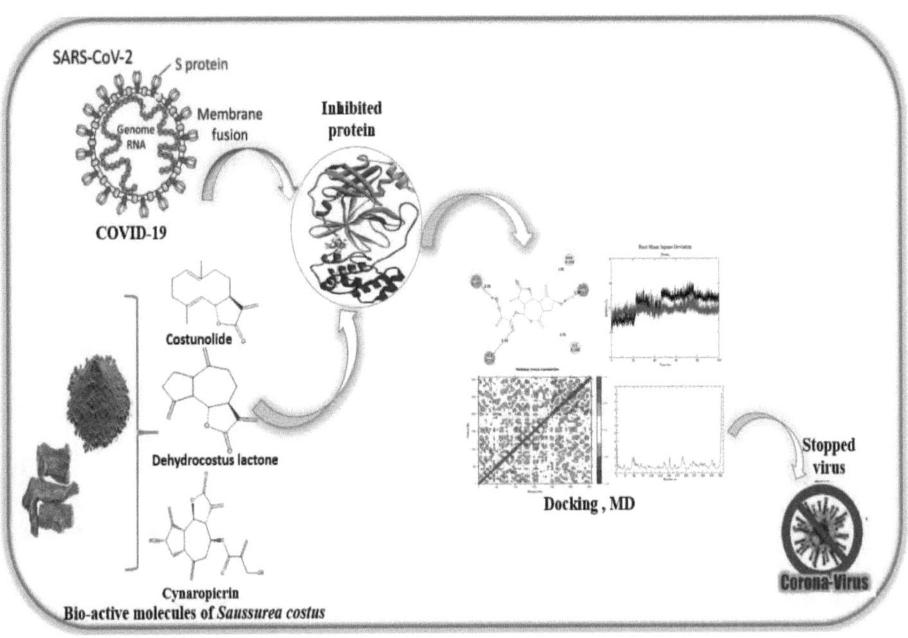

1. Introduction

Since (2019) COVID-19 virus has been announced a global pandemic, COVID-19 by the World Health Organization (W.H.O), on February 11th, 2020 stands for "Co": corona (meaning crown), "vi": virus, "d": disease, and "19": the year (2019) [1-5]. This virus which is labeled as (COV2) is different from both MERS-COV (Middle East Respiratory Syndrome-related Coronavirus) responsible for an epidemic that spread in 2012 in the Middle East [6], and SARS (Severe Acute Respiratory Syndrome Corona Virus1 that brought SARS epidemic about in 2003[7]. The virus enters the body through contact with the cells of the mucous membranes lining the nose, mouth, and eyes, causing infections in the respiratory system, and it takes at least six days (the incubation period) [8-9] for symptoms to appear in the infected patients. The main symptoms of COVID-19 disease are extreme fatigue, fever, sneezing, shortness of breath, coughing, and breathing difficulties. These symptoms are similar to those of a cold or the flu. Still, the impact of COVID-19 is different from one person to another, as the risks of severe complications of COVID-19 that led to death or prolonged recovery with weak lungs in smokers and elderly patients [10]. Zhang et al., examined the mechanism SARS-CoV-2 virus action with human body, it exploits the interplay between the small miRNA and other biomolecules to avoid being effectively recognized and attacked by host immune protection and deactivate functional genes crucial for the immune system. In detail, SARS-CoV-2 can be regarded as a sponge to adsorb host immune-related miRNA, which forces the host to fall into the dysfunctional status of the immune system. Besides, SARS-CoV-2 encodes its miRNAs, which can enter a host cell and are not perceived by the host's immune system, subsequently targeting host function genes to cause illnesses. Therefore, this article presents a reasonable view that the miRNA-based interplays between the host and SARS-CoV-2 may be the primary cause that SARS-CoV-2 accesses and attacks the host cells [35, 36].

During the epidemic, there was an increase in the consumption of *Saussurea costus* in Indonesia, which has a low rate of hospital admissions for Indonesian patients compared to other countries[11] . The Costus plant originated in India and has since spread to many countries worldwide. It has a ginger-like appearance, grows to a height of 1 meter and a half, and has been used for medicinal purposes since antiquity [12].

There are two types of Costus: the white, sweet "Marin" or "Bahri," which the Arabs brought to their countries by sea and whose taste is prized, and the black, bitter "Indian," which is thought to be superior in terms of medicine strength and defense against a wide range of diseases, especially influenza[13] . The most important part used in the treatment is the skin of the roots[14-15]

Costus powder is usually used for medical purposes, where the active ingredients are organic compounds. It has a pivotal role in treating allergies and is an antiseptic agent for germs. It is an essential antiseptic agent for tonsillitis germs, pharyngitis, bacterial pleurisy, colds, influenza, pneumonia, and several germs that infect the lungs. In addition, it has a significant role in shortness of breath, asthma, and cough. Because it prevents and mitigates harmful and sometimes fatal attacks of viruses on the respiratory system and supports the immune system. The oil and roots of Costus are the other parts of the plant used for medical purposes, which must be used scientifically and pharmacologically to avoid overdose poisoning [16].

Fig. 1. *Saussurea Costus* [17].

This research aims to investigate Costus plant active ingredients as a potential treatment for Covid-19 relying on the pharmaceutical properties of *Saussurea Costus*. The current study uses different computational approaches; such as Reverse docking, Molecular dynamics, and ADMET. In addition, Lipinski rules to screen the interest of this medicinal plant against SARS-CoV2 infection.

2. Computational details:

2.1. The bioactive molecules of the medicinal plant *Saussurea costus*:

Three organic bioactive compounds have led to increased interest in researchers, as shown in Figure 2 such as Costunolide, Dehydrocostus lactone and Cynaropicrin are sesquiterpene lactones found in *Saussurea costus* contains the most important amino acids, organic acids, essential oil, highlighting strong potential therapeutic effects helps to recover faster with various acute respiratory diseases, flu, sore throat and pleurisy [18-20].

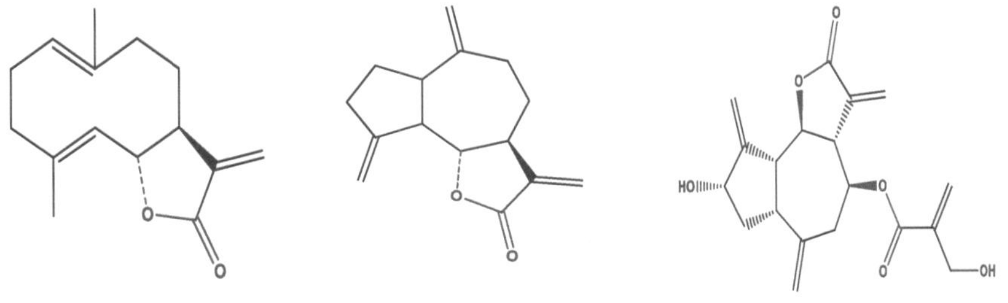

Costunolide   Dehydrocostus lactone   Cynaropicrin

Fig. 2. The targe bioactive molecules of *Saussurea costus*.

2.2. Reverse Docking:

Recently, the development of computer and bioinformatics programs has made great strides in predicting the molecular interactions that carry a protein and ligand at the binding site and in designing and discovering new drugs. The bioinformatics program that we applied in this research is reverse docking molecular modeling. This program is widely used mainly in medicinal chemistry and in designing new therapeutic molecules [21]. By fixing the target molecules at the disease protein's binding sites and looking for the most stable interactions, reverse docking aids in the discovery of the best position for the bioactive molecule and disease protein. The molecular structures of Costunolide, Dehydrocostus and Cynaropicrin were energetically minimized and geometrically optimized by using Gaussian software. Then the realization of reverse docking is also based on the preparation of proteins : The crystal structures of SARS-CoV-2 are available in the Protein Data Bank (PDB, https://www.rcsb.org/), the five models for high-resolution SARS-CoV-2 (PDB ID: 2GZ9 (2.17 Å); 6LU7 (2.16 Å); 7AOL (1.47 Å), 6Y2E (1.75 Å), 6Y84 (1 .39 Å) were downloaded from the Protein Data Bank Figure 3. These proteins were prepared by removing water molecules, adding polar hydrogens and Kohlman charges by using the Discovery Studio program, the prepared files are then converted to PDB format. As well as the sybyl surflex-dock software is used for the calculations and the study of molecular docking. The studied complexes were analyzed and reassembled in PDB format by using PyMol software [22-23]. Finally they were sent to the Discovery Studio software to visualize the interactions [24].

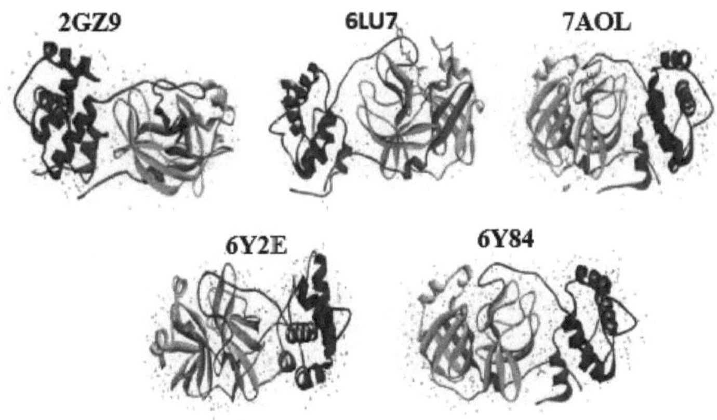

Fig. 3. 3-Dimentional structures of SARS-CoV-2 target proteins.

2.3. Molecular Dynamics Simulations:

GROMACS simulation package (GROMACS 2020.4) used to perform molecular dynamics simulations for protein-ligand complex was carried out for 100 ns in water using CHARMM36 force field; trajectory and energy files written every 10 ps [25].

The system was solvated in a truncated octahedral box containing TIP3P water molecules. The protein was centered in the simulation box within a minimum distance to the box edge of 1 nm to satisfy the minimum image convention efficiently. The simulation was performed in 0.15M KCl by adding 60 Potassium ions and 57 Chloride ions, and the overall system contained 66242 atoms.

Minimization was carried out for 5000 steps using Steepest Descent Method, and the convergence was achieved within the maximum force < 1000 (KJ mol^{-1} nm^{-1})

to remove any steric clashes. All three systems were equilibrated at NVT and NPT ensembles for 100ps (50,000 steps) and 1000ps (1,000,000 steps), respectively, using time steps 0.2 and 0.1 fs, at 300K to ensure a fully converged system for a production run.

The production runs for simulation were carried out at a constant temperature of 300 K and a pressure of 1 atm or bar (NPT) using weak coupling velocity-rescaling (modified Berendsen thermostat) Parrinello-Rahman algorithms, respectively. Relaxation times were set to $\tau T = 0.1$ ps and $\tau P = 2.0$ ps. All bond lengths involving hydrogen atoms were kept rigid at ideal bond lengths using the Linear Constraint Solver (lincs) algorithm, allowing for a time step of 2 fs. Verlet scheme was used for the calculation of non-bonded interactions. Periodic Boundary Conditions (PBC) were used in all x, y, z directions. Interactions within a short-range cutoff of 1.2 nm were calculated in each time step. Particle Mesh Ewald (PME) was used to calculate the electrostatic interactions and forces to account for a homogeneous medium outside the long-range cutoff. The production was run for 100ns for the complex [26].

2.4. Predicted pharmacokinetic and toxicity properties

It is the science of drug kinetics Absorption, Distribution, metabolism, and elimination (kinetic study of ADMET), [A]bsorption the passage of the drug from the site of administration into the general circulation. [D]istribution is the movement of blood to and from the tissues. [M]etabolism of drugs is the chemical modification of the drug with the overall goal of getting rid of the drug; enzymes are generally involved in metabolism. [E]limination is the irreversible removal of the parent drug from the body. [T]oxicity as the name suggests, this filter is used to measure the value of a compound and its metabolites [27-29].

Lipinski's rules are a list of guidelines proposed by Chris Lipinski of Pfizer to help design molecules with good availability. The four guidelines are: (1) the molecule must have a molecular weight MW of 500g/Mol or less. (2) The molecule should have no more than 5 hydrogen bond donors HBDs. These are mainly OH and NH groups. (3) Molecules should have no more than 10 HBAs than a hydrogen bond acceptor. They are essentially the oxygen and nitrogen atoms in the molecule (4). The lipophilicity of the molecule, measured by logP, must not be greater than 5. These have been called "Lipinski's Rules" or "Rule of 5" [30-31]. Lipinski recommended that the molecules adhere to at least three of the four guidelines to have a reasonable chance of having beneficial oral bioavailability. Therefore, when we think about ADME, Lipinski's rules help meet the letter A-absorption or have higher efficacy or lower toxicity. Therefore, these rules focus only on improving the solubility and membrane permeability to maximize oral absorption.

Computational approaches can help minimize the risks of toxicity. In this research, a new approach pkCSM have developed. It is considerably better than current methods, where the freely accessible web servers (SwissADME software) that Simplified Molecular Input Line System (SMILES) used during the generation process. [32].

3. Results and discussions:

3-1. Reverse Docking:

Reverse docking was applied on three molecules of *Saussurea costus* Costunolide, Cynaropicrin, and Dehydrocostus Lactone to identify therapeutic activity against Sarc-Cov-2 [18]. It is a technique that allows predicting or studying the interactions that are the main factors that significantly impact the affinity of a ligand for a receptor. In addition, this study helps us choose the effective protein for some compounds based on *Saussurea costus*, considered a treatment against COVID-19. The five proteins that are available in the Protein Data Bank with good resolution (ID PDB: 2GZ9 (2,17 Å); 6LU7 (2,16 Å); 7AOL (1,47 Å), 6Y2E (1,75 Å), 6Y84 (1,39 Å) on three molecules of Saussurea costus Costunolide, Cynaropicrin, and Dehydrocostus as shown in the Figure 4. For PDB 2GZ9 (2.17 Å) (Figure 5), it was observed that there are just Pi-Alkyl-type interactions with the residues PHE 8 and PHE 294 in the case of Costunolide. On the other hand, visualization of docking of Cynaropicrin docked with the same protein give two best binding modes of protein with green color for Amino acid residues involved in hydrogen bonding interaction with residues THR 111 and TYR 154, As well, this complex provides another interaction Pi-Alkyl with residue PHE 294. Moreover, Dehydrocostus lactone shows Carbon-Hydrogen Bond interactions with residues PRO 108, Pi-Alkyl with ILE 200, LEU 202, and HIS 246. In addition, the PDB: 6LU7 complex (Resolution: 2.16 Å) (Figure 6) with Cynaropicrin causes hydrogen bonds with the OH groups, which show the importance of these amino acids THR 111, ASP 153, and SER 158, which are considered among the essential factors of stability protein-ligand. Moreover, the distances of its bonds are lower than 4Å, which is confirmed by the strong hydrogen bonds; they are 1.56 Å, 2 Å and 3.34 Å respectively. Regarding 6LU7 protein with these two molecules, Costunolide and Dehydrocostus Lactone made Pi-Alkyl interactions with phenylalanine. On the one hand, the presence of hydrogen bond interaction of complex 7AOL (Resolution: 1.47 Å) (Figure7) and Costunolide with residue LYS 97 at a distance of 2.60 Å, a Carbon Hydrogen Bond interaction with ALA 70 at a distance of 2.52 Å and two Alkyl interactions with TRP 31 and VAL18 at distances of 4.46Å and 5.01 Å, respectively. Thus, with Cynaropicrin, there are two interactions: Conventional Hydrogen Bond and Carbon-Hydrogen Bond with the amino acids GLN 110, SER 158 successively. Dehydrocostus Lactone is associated with the presence of two interactions formed with the amino acids of 7AOL, Pi-Alkyl with PHE 8 at a distance of 4.86 Å, and Pi-Donor Hydrogen Bond with PHE 294 at distances 3.26 Å.

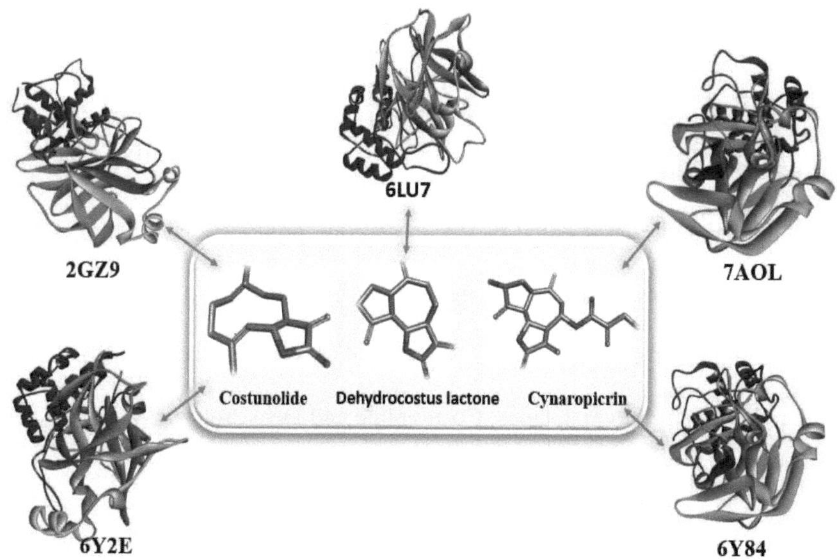

Fig 4. Illustration of costunolide, dehydrocostus lactone and cynaropicrin inhibitions against the SARS-CoV-2 target proteins : 2GZ9, 6LU7, 7AOL, 6Y2E, and 6Y84

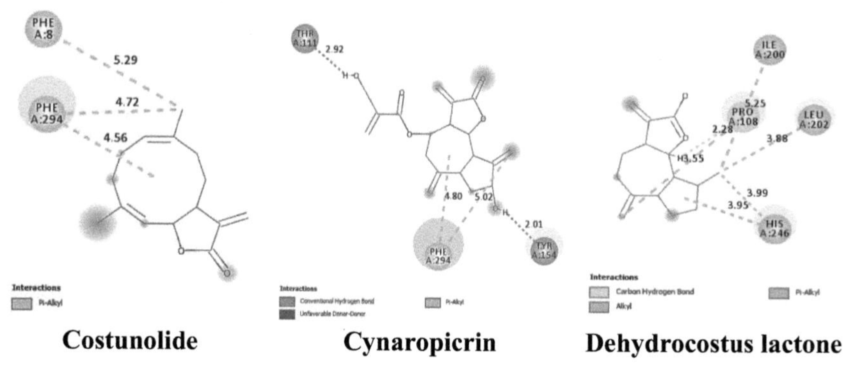

Fig 5. Docking interactions of costunolide, dehydrocostus lactone and cynaropicrin (**PDB :2GZ9**)

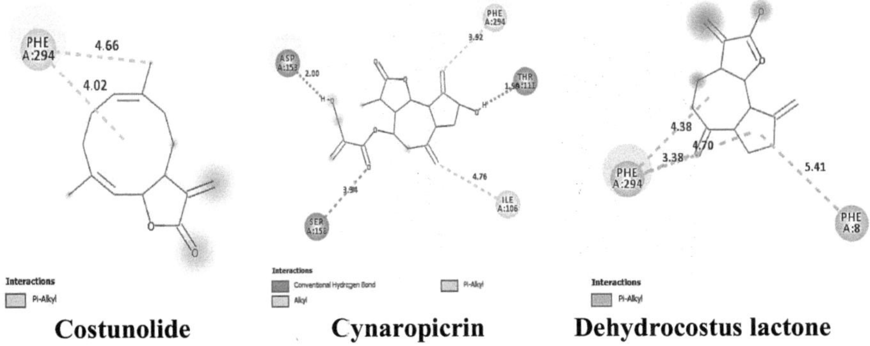

Costunolide　　　**Cynaropicrin**　　　**Dehydrocostus lactone**

Fig 6. Docking interactions of costunolide, dehydrocostus lactone and cynaropicrin (**PDB : 6LU7**)

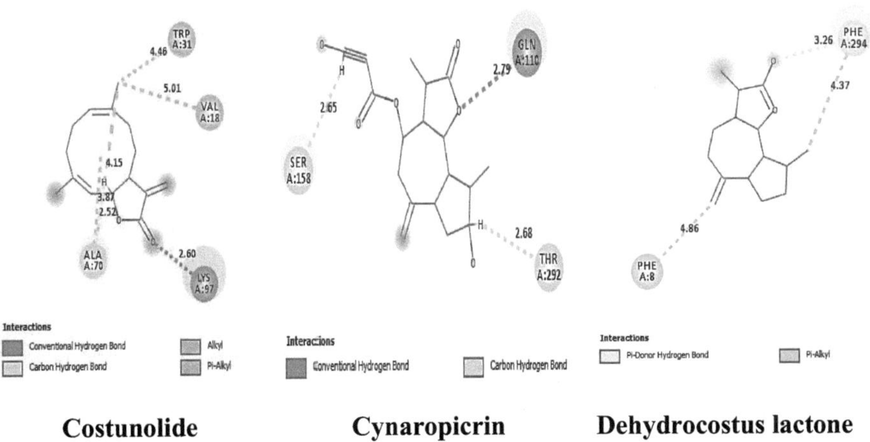

Costunolide　　　**Cynaropicrin**　　　**Dehydrocostus lactone**

Fig 7. Docking interactions of costunolide, dehydrocostus lactone and cynaropicrin (**PDB : 7AOL**)

Table 1. Different interactions and key residues (amino acids) for more active molecules.

Bioactive molecules	Target proteins	Amine acid	Position	Distance	Interaction type	Total- score
Costunolide		PHE	8	5.29	Pi-Alkyl	2,34
		PHE	294	4.56		
Cynaropicrin		THR	111	2.92	Conventional Hydrogen Bond	3,45
		TYR	154	2.01		
		PHE	294	4 .80	Pi-Alkyl	
Dehydrocostus lactone	2GZ9	ILE	200	5.25	Pi-Alkyl Alkyl	1,89
		LEU	202	3.88		
		HIS	246	3.95		
		PRO	108	2.28	Carbon Hydrogen Bond	
Costunolide		PHE	294	4.02	Pi-Alkyl	1,69
Cynaropicrin		THR	111	1.56	Conventional Hydrogen Bond	4,78
		ASP	153	2.00		
	6LU7	SER	158	3.34		
		PHE	294	3 .92	Pi-Alkyl Alkyl	
		ILE	106	4.76		
Dehydrocostus lactone		PHE	294	3.38	Pi-Alkyl	2,03
		PHE	8	5.41		
Costunolide		LYS	97	2.60	Conventional Hydrogen Bond	2,36
		TRP	31	4.46	Pi-Alkyl Alkyl	
		VAL	18	5.01		
		ALA	70	2.52	Carbon Hydrogen Bond	
Cynaropicrin	7AOL	GLN	110	2.79	Conventional Hydrogen Bond	4,36
		SER	158	2.65	Carbon Hydrogen Bond	
Dehydrocostus Lactone		PHE	8	4.86	Pi-Alkyl	1,55
		PHE	294	3.26	Pi-Donor Hydrogen Bond	

On the other hand, the reverse docking study for the rest of the proteins, such as 6Y2E (Resolution 1.75 Å) and 6Y84 (Resolution 1.39 Å) (Figure 8&9), give unsatisfactory results for Cynaropicrin. Also, it showed the presence of unfavourable donor-donor interactions with the amino acids LYS 102 and PHE 294, respectively

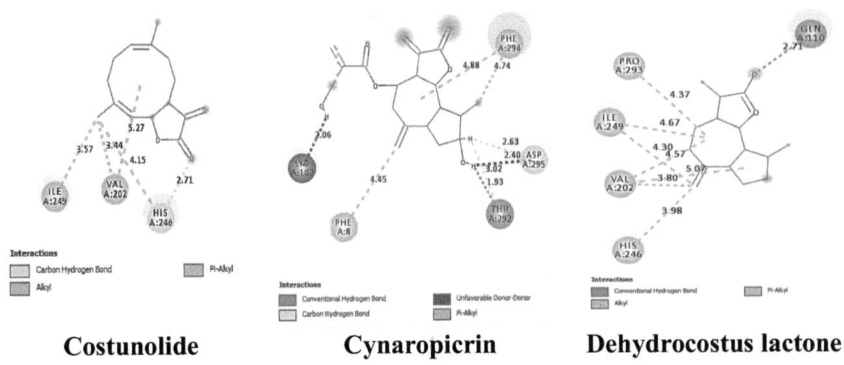

Costunolide   Cynaropicrin   Dehydrocostus lactone

Fig. 8. Docking interactions of costunolide, dehydrocostus lactone and cynaropicrin (**PDB : 6Y2E**)

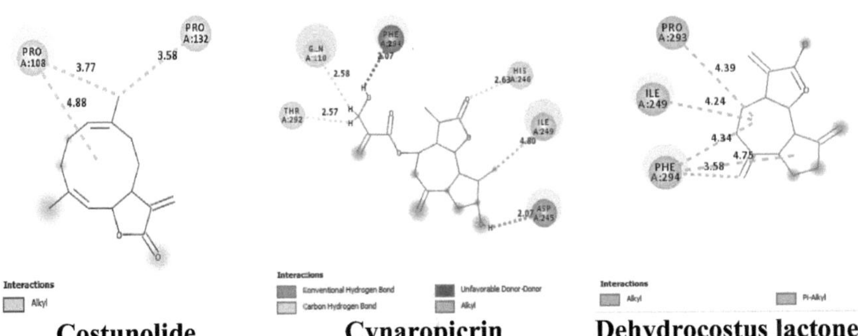

Costunolide   Cynaropicrin   Dehydrocostus lactone

Fig. 9. Docking interactions of costunolide, dehydrocostus lactone and cynaropicrin (**PDB : 6Y84**)

Table 2. Different interactions and key residues (amino acids) for inactive molecules.

Bioactive molecules	Target proteins	Amine Acid	Position	Distance	Interaction type	Total score
Costunolide	6Y2E	ILE	249	3.57	Pi-Alkyl	1,62
		VAL	202	3.44	Alkyl	
		HIS	246	2.71	Carbon Hydrogen Bond	
Cynaropicrin		THR	292	1.93	Conventional Hydrogen Bond	4,20
		PHE	294	4 .74	Pi-Alkyl	
		PHE	8	4.45		
		LYS	102	2.06	Unfavorable Donor-Donor	
Dehydrocostus lactone		GLN	110	2.71	Conventionnel Hydrogen Bond	1,41
		ILE	249	4.30	Pi-Alkyl Alkyl	
		VAL	202	3.80		
		HIS	246	3.98		
		PRO	293	4.37	Carbon Hydrogen Bond	
Costunolide	6Y84	PRO	108	3.77	Pi-Alkyl	1,56
		PRO	132	3.5		
Cynaropicrin		ASP	245	2.07	Conventional Hydrogen Bond	2,53
		ILE	249	4 .80	Alkyl	
		GLN	110	2.58	Carbon Hydrogen Bond	
		THR	292	2.57		
		HIS	246	2.63		
		PHE	294	2.07	Unfavorable Donor-Donor	
Dehydrocostus lactone		ILE	249	4.24	Pi-Alkyl Alkyl	2,24
		PRO	293	4.39		
		PHE	294	3.58		

According to molecular docking studies, the highest total score strongly denotes

strong binding affinity. On the other hand, the molecule which will have minimum

energy is considered the good one, so according to the two tables 1&2, the best

complex is [6LU7-Cynaropicrin] because it has a high score of 4.78 and low

energy compared to the others with a value of -7.4 kcal/mol and is not the only

criteria, as well the best complex is [6LU7-Cynaropicrin] because of its chemical

structure and docking interactions, as shown in (Figure 10) is a guaianolide type

known as a sesquiterpene lactone. It has a 5-7-5 fused tricyclic skeleton with four

exo-olefins, six stereocenters, and two hydroxyl groups, which are caused

meaningful interactions of conventional Hydrogen Bond with THR 111, ASP 153

residues. Also, Pi-Alkyl and Alkyl interact with the amino acids PHE294 and ILE

106. Using molecular dynamics simulation, the structure of the SARS-CoV-2

main protease (PDB ID 6LU7) complex with the inhibitor cynaropicrin to confirm

its stability.

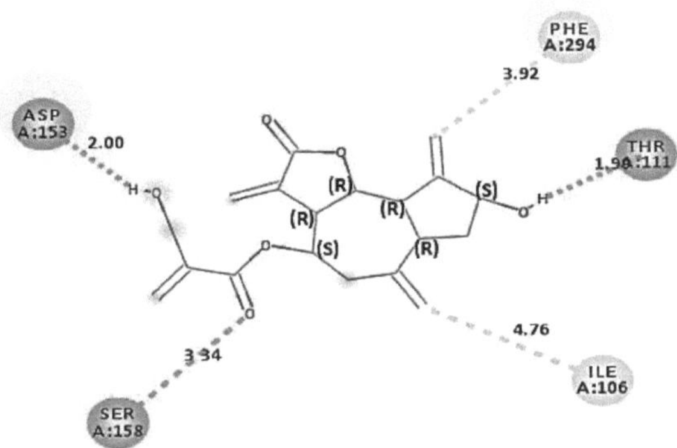

Fig. 10: Cynaropicrin structure with different interactions.

3-2. Validation of Docking Protocol :

The molecular docking validation step is crucial. In this research, the re-docking method was used, where a reference ligand is downloaded with the target protein to PYMOL software, which forms a new complex (docked) [37-40]. Fig. 11 demonstrates that the two ligands are positioned identically. In this operation, the small RMSD value of 1.986 Å indicates the accuracy of the docking results, which is close to the crystallography results.

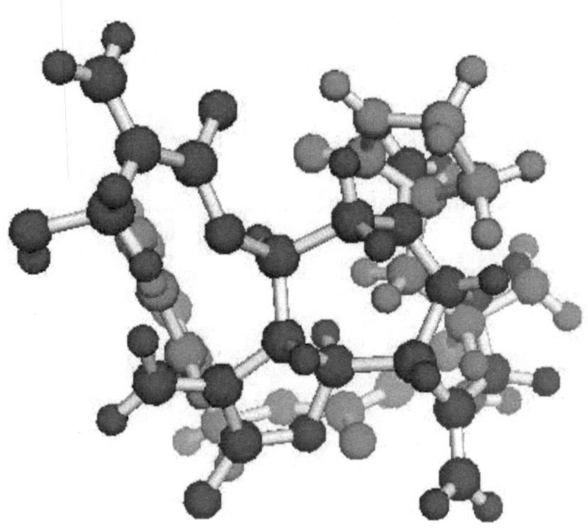

Fig. 11. Overlays of Re-docking pose and RMSD value of 1.986 Å
(Original = Green, Docked=red) visualized with PyMOL

3-3. Molecular Dynamics Simulations :

3-3-1. RMSD (Protein)

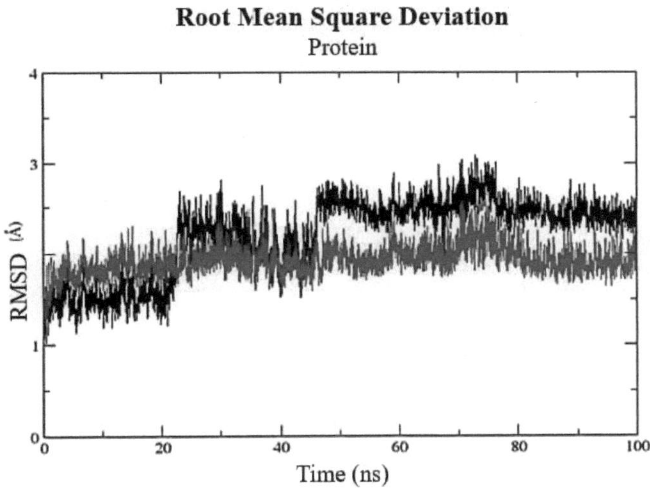

Fig. 12. RMSD for Cynaropicrin-6LU7 complex based on "C-alpha" atoms. Fluctuation within a range of ~1.5 Å (in black color line) and the (red color line) after removing the last 6 residues of C-terminal.

RMSD calculated the complex based on 'C-alpha' atoms using gromacs program. The C-terminal of the protein appears to be highly flexible (as indicated in RMSF), due to which the RMSD shows some fluctuation within a range of ~1.5 Å (in black). RMSD of protein (in Red) after removing the last 6 residues of C-terminal to check their effects on RMSD graph. The RMSD plot in red shows a relatively stable protein. The mean RMSD value for protein in black is 2.19 ± 0.43 Å and for protein in Red is 1.91 ± 0.15 Å.

3-3-2. RMSD (Ligand)

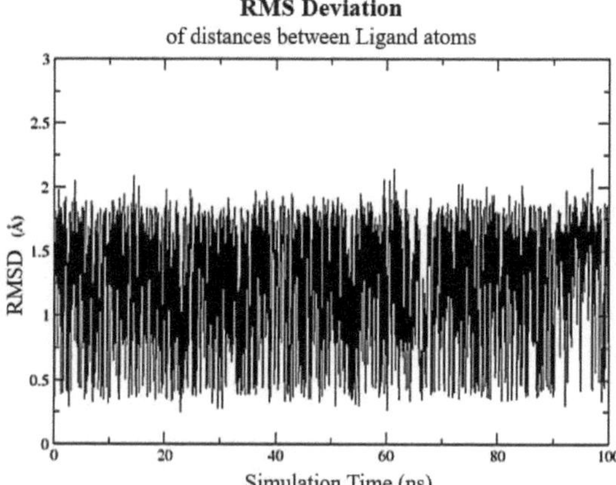

Fig. 13. RMSD for Cynaropicrin-6LU7 complex based on ligand's atoms
RMSD calculated for the ligand-based on ligand's atoms using gromacs
program.
The mean RMSD value is 1.25 ± 0.42 Å. RMSD of the ligand shows high

fluctuations of the ligand. It also indicates that ligand exists most of the time in

free form and not in bound form.

3-3-3. RMSF

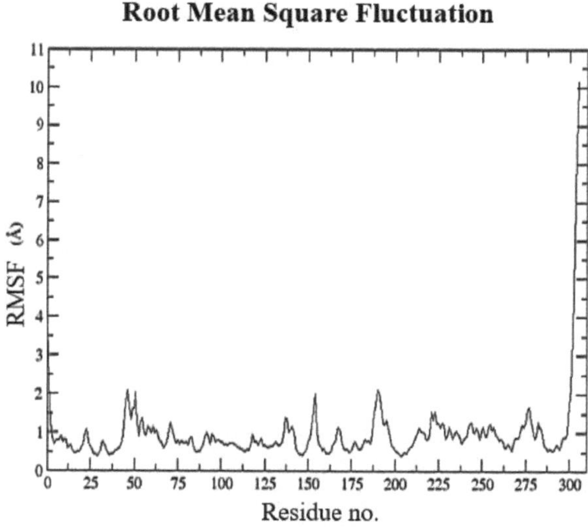

Fig. 14. RMSF calculated for Cynaropicrin based on "C-anlpha" atoms.

A RMSF was calculated for the protein complex based on 'C-alpha' atoms using gromacs program. The whole protein remained relatively stable except for the C-terminal, which showed significant fluctuation.

3-3-4. ROG

Radius of Gyration

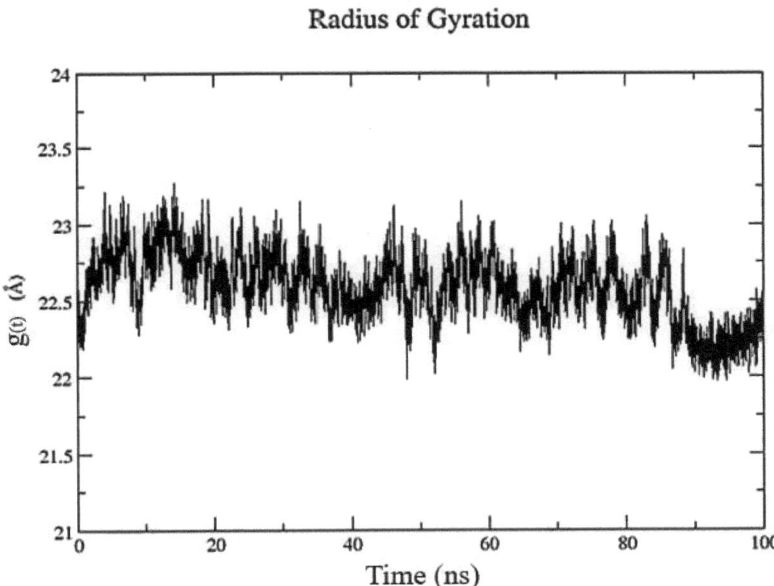

Fig.15. ROG calculated =**22.58 ± 0.22** Å

Radius of gyration calculated for the complex based on 'C-alpha' atoms using the gromacs program. The mean RoG value is 22.58 ± 0.22 Å. A very slight fluctuation is observed, indicating a slight opening and closing of protein. C-terminal plays a significant role in this fluctuation. Overall, the structure remains stable.

3-3-5. Protein-Ligand complex

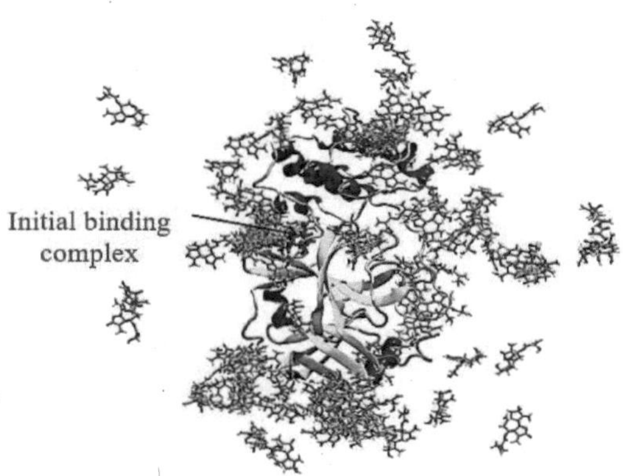

Fig. 16. Complex during 100 ns of simulation time.

100 snapshots of ligand taken at every 1ns during 100 ns of simulation time. The image indicates that the ligand does not bind with the protein. The initial conformation of ligand (0ns) is shown is red licorice. The conformation of protein is set to the initial conformation referring to 0ns of simulation.

3-3-6. Hydrogen Bonds (Protein-ligand)

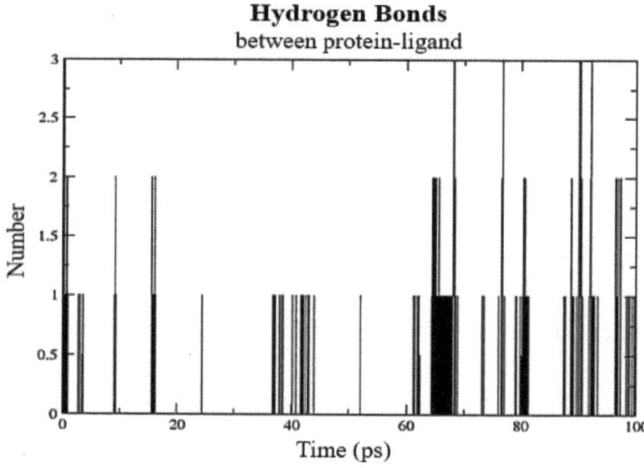

Fig. 17. Total number of hydrogen bonds formed between ligand and protein during 100ns of simulation time. In-consistent fluctuating h-bonds indicate that ligand has left the protein several times.

3-3-7. Center of mass distance between Ligand and Protein

Average distance

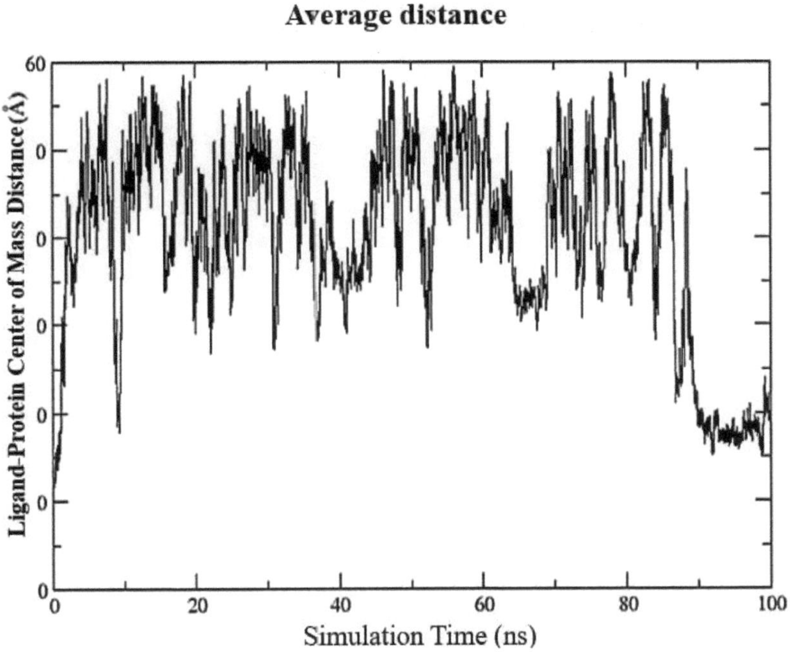

Fig. 18. Average Center-of-Mass Distance between ligand and protein during 100ns of simulation time. The mean distance is **40.24 ± 11.08** Å.

The plot indicates that the ligand leaves the binding site and roams around the protein, with multiple short interactions. The more considerable COM distance continues until 90 ns, and later it comes closer to protein.

3-3-8. Principal Component Analysis

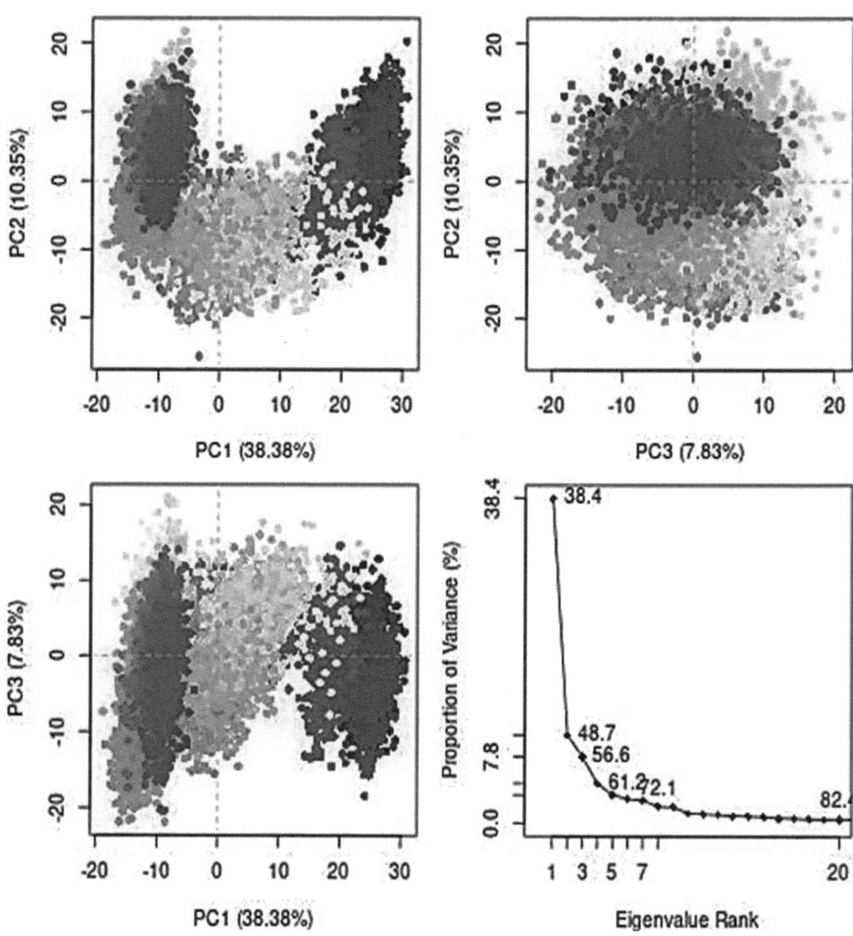

Fig. 19. Principal Component analysis of the complex calculated from Bio3D program of R. All three PCs captured 56.6% of structural variance in protein.

3-3-9. PCA-1 and PCA-2 of protein

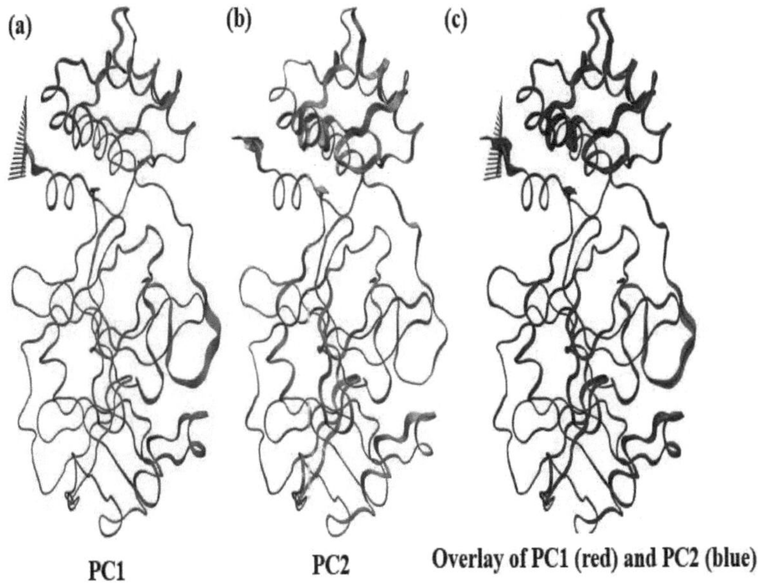

Fig. 20. Interpolated structures of protein along (a) PC1, (b) PC2, and (c) the overlay conformations.

In this view, atoms are colored on a scale from blue to red, where blue represents atoms showing large motion amplitudes, and red represents more rigid atoms. The(c) shows the overlay conformations of proteins along both PC1 (shown as red) and PC2 (shown as blue), which are orthogonal to each other.

3-3-10. Dynamic Cross Correlation Matrix Analysis (DCCM)

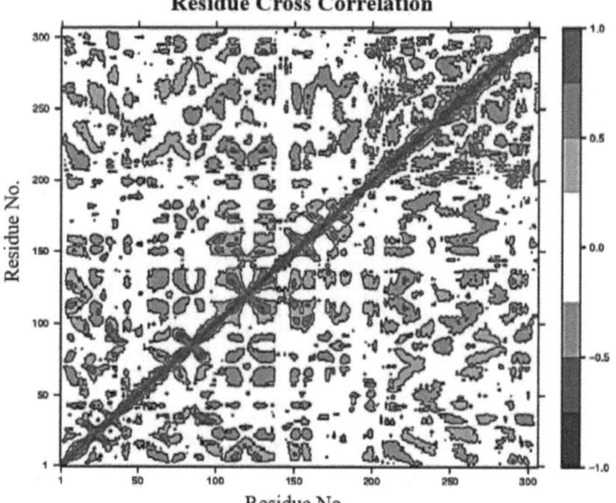

Fig. 21. Protein Residue dynamic cross-correlated motions for the complex calculated from Bio3D program of R.

Colors varying from red to white to blue indicate the intensity of correlated motion. Blue colors indicate negative correlation, white shows no correlation, and pink shows positively correlated motions between residues.

3-3-11. Potential energy

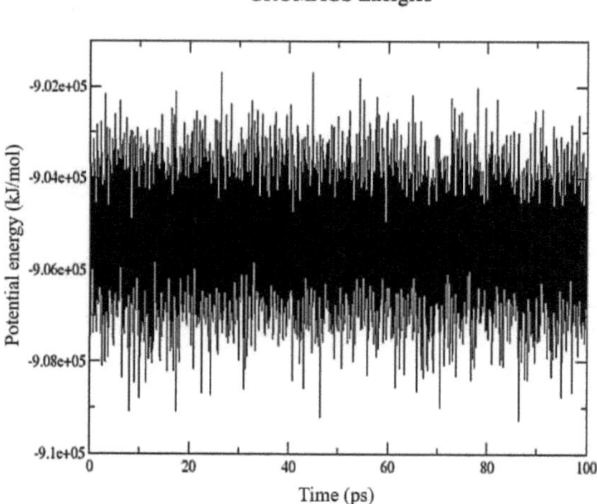

Fig. 22. Potential energy of system during 100 ns.

The system's potential energy during 100 ns of MD simulation as obtained from Gromacs energy file. The graph shows the converged potential energy.

3-3-12. Pressure

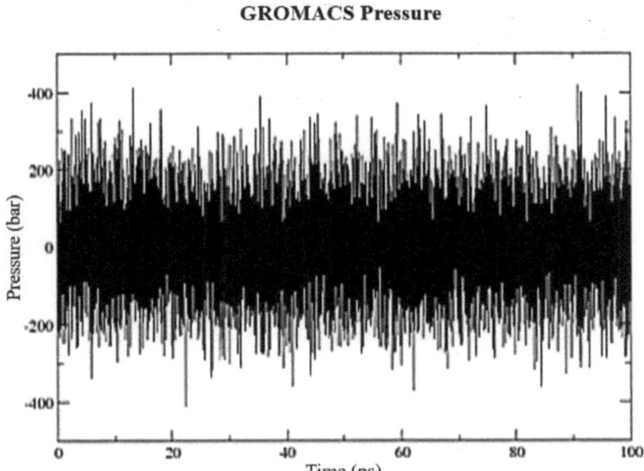

Fig. 23. Total pressure of system during 100 ns of MD simulation as obtained from Gromacs energy file.

3-3-13. Temperature

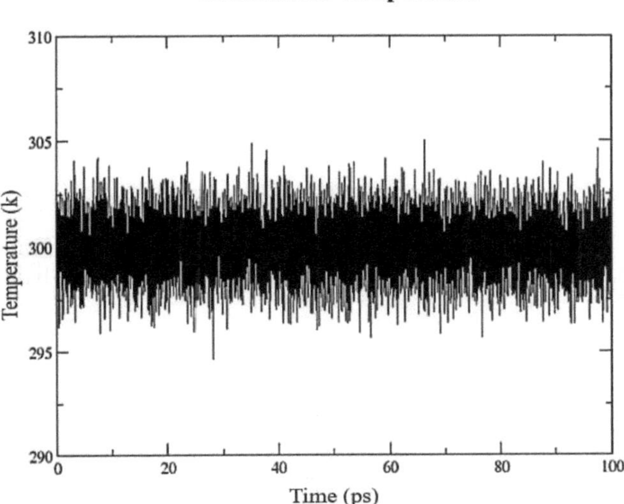

Fig.24. Temperature of system during 100 ns of MD simulation as obtained from Gromacs energy file.

3-3-14. MMGBSA Binding Energy

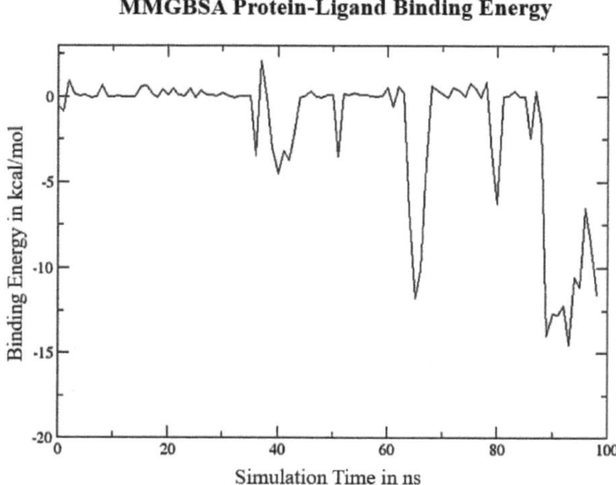

Fig. 25. Binding energy of complex during 100 ns of MD simulation.

The binding energy of ligand-protein complexes during 100 ns of MD simulation was obtained by the MMGBSA method, where the value was -1.74 ± 3.92 kcal/mol. The binding free energy is a crucial factor in determining the activity of drug molecules, and the lower value of ΔG_{bind} value, the more stable Enzyme-molecule complex formed.

Table 3. MMGBSA binding energy in kcal/mol for protein-ligand [6LU7-Cynaropicrin] complex.

6lu7 Complex with:	ΔE^{VDW} (van der Waal's energy)	ΔE^{elec} (Coulombic energy)	ΔG^{GB} (Generalized-Born Polar solvation energy)	ΔE^{SASA} (Non-Polar solvation energy)	ΔG^{MMGBSA} (Protein-Ligand Binding energy)
complex	-3.69 ± 6.35	-1.80 ± 4.53	4.26 ± 7.19	-0.49 ± 0.87	-1.74 ± 3.92

4. ADMET

The discovery of a new molecule to be a potential drug requires an investigation of pharmacokinetic properties. Hence, the evaluation of ADMET properties is considered the best method to check and choose the target molecules via in silico methods using the online services available at Pkcsm.

Absorption :

A review of the literature indicated that molecules or drugs with results (>30% abs) have more potency in crossing the intestinal barrier [33]. Consequently, the results show that all the compounds studied showed high values; the percentage of human oral absorption varied between 84.54 and 97.607%, indicating that these compounds were considered well absorbed.

Distribution :

Distribution values predicted that the bioactive molecule of interest, Cynaropicrin has a low value of -0.47 in Blood-Brain Barrier Permeability, which designates that don't enter the brain. Thus, undesirable effects will be reduced, proving their efficacy and drug-likeness behavior.

Metabolism :

Computational metabolic behavior indicates the chemical biotransformation of a drug in the body; CYP plays a significant role in converting drug compounds. In

Table 4, Cynaropicrin compound is revealed as substrates and inhibitors of the two main subtypes, 2D6 and 3A4 of CYP, meaning they may have the ability to be metabolized in the liver.

Excretion and toxicity :

The excretion result on Cynaropicrin shows an acceptable negative value; it is then evaluated for the Ames toxicity test, which is considered non-toxic; in parallel, in terms of Hepatotoxicity, Cynaropicrin has proven to be non-toxic, which justifies its effectiveness.

Lipinski's rule:

The results of the toxicity tests of the ADMET method have proven that the bioactive molecule which interests us, Cynaropicrin, is the least toxic and the safest possible. Moreover, it can be observed from Table 5 that these three compounds comply with Lipinski's rule standards: they formed ≤5 H-bond donors and ≤ 10 H bond acceptors, MW < 500 Daltons, and octanol/water partition coefficient < 5. The ADMET analysis and Lipinski's rules indicate that the designed Cynaropicrin was verified in silico as a safe pharmaceutical compound.

Table 4: the results of the ADMET test with pKCSM of all compounds.

	Absorption	Distribution		Metabolism				Excretion	Toxicity	
		Blood Brain Barrier Permeability	CNS permeability	CYP				Renal OCT2 substrate	AMES toxicity	Hepatotoxicity
	Intestinal absorption (human)			2D6	3A4	2D6	3A4			
				Substrate		Inhibitor				
	Numeric (% Absorbed)	Numeric (log BB)	Numeric (log PS)	Categorical (Yes/No)				Categorical (Yes/No)	Categorical (Yes/No)	
Costunolide	97.026	0.511	-2.69	No	No	No	No	No	No	No
Cynaropicrin	84.54	-0.47	-2.984	No	No	No	No	No	No	No
Dehydrocostus lactone	97.607	0.596	-2.184	No	Yes	No	Yes	Yes	No	No

5. Lipinski Rule of five (RO5) and Veber rule :

Table 5 : Lipinski's and Veber Rules parameters of all compounds in the dataset

Compounds	Lipinski Rules				Veber Rules			
	HBD*	HBA*	MW	Log P	t-PSA	RB	HBA	HBD
Costunolide	0	2	232.323	3.55	26.3	0	1	0
Cynaropicrin	2	6	346.379	1.05	93.06	3	4	2
Dehydrocostus Lactone	0	2	230.307	3.01	26.30	0	1	0
Reference ligand	≤ 10	≤ 5	150 to 500	-0,7 to +5	≤ 140	≤ 9	≤ 5	≤ 10

HBD : number of Hydrogen Bond Donors (*according to the Lipinski rule) ; HBA : Hydrogen Bond acceptors (*according to the Lipinski rule) ; RB :number of Rotatble Bonds ; TPSA :Total Polar Surface Area, MW : molecular weight ; logP :the llog of octanol-water parition coefficient

.

The compounds showed drug-like characteristics based on the rules of Lipinski and Veber (Table 5), which meant that they had a higher probability of good absorption after oral administration. According to Lipinski's rule, Veber and ghose these molecules exhibited drug-like properties.

Conclusion :

Traditional therapeutic medicine is considered a real assistant for human health and well-being. In this research three bioactive molecules of *Saussurea costus* Costunolide, Cynaropicrin and Dehydrocostus Lactone are studied and discovered as a potential medication for COVID-19. The reverse docking was conducted to determine the inhibitory action of these compounds' constituents with the different protease of COVID-19. According to the visualization results of the interactions of all the complexes the best complex is of [6LU7-Cynaropicrin]. Furthermore, molecular dynamics simulations confirmed [6LU7-Cynaropicrin] complex stability. Finally, the results of Lipinski's Rule for [6LU7-Cynaropicrin] complex are very satisfactory, and it could be considered as a potential drug for COVID-19. In addition, Cynaropicrin is a small molecule that can be easily synthesized therefore the synthesis of Cynaropicrin and the examination of their *in vitro* activity against SARS-CoV-2 could be interesting and help to obtain new effective drugs.

References

[1] M.A. Hitt, R.M. Holmes, J-L. Arregle, The (COVID-19) pandemic and the new world (dis)order, J. World Bus. 56:4 (2021) 101210. https://doi.org/10.1016/j.jwb.2021.101210.

[2] G. Gnanasegaran, D. Paez, M. Sathekge, F. Giammarile, S. Fanti, A. Chiti, H. Bom, S. Vinjamuri, T.N.B. Pascual, J. Bomanj, Coronavirus (COVID-19) pandemic mediated changing trends in nuclear medicine education and training: time to change and scintillate, Eur J Nucl Med Mol Imaging 49:2 (2022) 427–435. https://doi.org/10.1007/s00259-021-05241-2.

[3] A.M. Mokhtari, R. S. Dewey, A. Mirahmadizadeh. The Global Challenges of Controlling Coronavirus Disease 2019: A Review Study, J. Health Sci. Surveill. Syst. 9:3 (2021) 142-148. 10.30476/JHSSS.2021.90136.1179.

[4] N. Singh, Y. Tang, Z. Zhang, C. Zheng, COVID-19 waste management: Effective and successful measures in Wuhan, China, Resour Conserv Recycl. 163 (2020) 105071. https://doi.org/10.1016/j.resconrec.2020.105071.

[5] S.E. Park, Epidemiology, virology, and clinical features of severe acute respiratory syndrome -coronavirus-2 (SARS-CoV-2; Coronavirus Disease-19), Clin Exp Pediatr. 63:4 (2020) 119–124. https://doi.org/10.3345/cep.2020.00493.

[6] H. Wang, J.E. Stokes, J.A. Burr, Depression and Elevated Inflammation Among Chinese Older Adults: Eight Years After the 2003 SARS Epidemic, The Gerontologist. 61:2 (2021) 273–283. https://doi.org/10.1093/geront/gnaa219

[7] T.G. Ksiazek, D. Erdman, C.S. Goldsmith, S.R. Zaki, T. Peret, S. Emery, et al. A Novel Coronavirus Associated with Severe Acute Respiratory Syndrome, N Engl J Med. 348 (2003) 1953-1966. DOI: 10.1056/NEJMoa030781.

[8] L.M. Weng, X. Su, X.Q. Wang, Pain Symptoms in Patients with Coronavirus Disease (COVID-19): A Literature Review, J Pain Res. 26:14 (2021) 147- 159. DOI: 10.2147/JPR.S269206

[9] N. Magnavita, G. Tripepi, R.R. Di Prinzio, Symptoms in Health Care Workers during the COVID-19 Epidemic. A Cross-Sectional Survey, Int J Environ Res Public Health. 17:14 (2020) 5218. doi: 10.3390/ijerph17145218

[10] R. Gyawali, P.N. Paudel, D. Basyal, W.N. Setzer, S. Lamichhane, M.K. Paudel, S. Gyawali, K. Prajwal, A Review on Ayurvedic Medicinal Herbs as Remedial Perspective for COVID-19, JKAHS. 3 (2020) 1-21. https://jkahs.org.np/jkahs/index.php/jkahs/article/view/237

[11] S.R. Prawiro, K. Anam, B. Prabowo, R. Bramanthi, A.A. Fitrianingsih, D.Y.N. Hidayati, S. Imawati, E. Fitria, S. Winarsih, Generating the Responses Immune with Honey, Saussurea costus, and Nigella Sativa in Cellular and Humoral May Resolve COVID-19, Sys Rev Pharm. 12(1) (2021) 1588-1593. doi:10.31838/srp.2021.1.224

[12] K. Zahara, S. Tabassum, S. Sabir, M. Arshad, R. Qureshi, M.S. Amjad, S.K. Chaudhari, A review of therapeutic potential of Saussurea lappa-An endangered plant from Himalaya, Asian Pac J Trop Med. 7 (2014) S60-S69. https://doi.org/10.1016/S1995-7645(14)60204-2

[13] P.V. Prasad. General medicine in Atharvaveda with special reference to Yaksma (consumption/tuberculosis), Bull Indian Inst Hist Med Hyderabad. 32(1) (2002) 1-14.

[14] M. Deabes, S. Abd-El Fatah, S. Salem, K. Naguib, Antimicrobial activity of bioactive compounds extract from Saussurea costus against food spoilage microorganisms, Egypt J Chem. 64:6 (2021) 2833-2843. https://doi.org/10.21608/ejchem.2021.69572.3528.

[15] S.I. Abdelwahab, M.M.E. Taha, H.A. Alhazmi, W. Ahsan, Z. Rehman, M. Al Bratty, H. Makeen, Phytochemical profiling of Costus (Saussurea lappa Clarke) root essential oil, and its antimicrobial and toxicological effects, Trop. J. Pharm. Res. 18: 10 (2019) 2155-2160. https://doi.org/10.1016/S1995-7645(14)60204-2

[16] M.A. El Hassab, A.A. Shoun, S.T. Al-Rashood, T. Al-Warhi, W.M. Eldehna, Identification of a New Potential SARS-COV-2 RNA-Dependent RNA Polymerase Inhibitor via Combining Fragment-Based Drug Design, Docking, Molecular Dynamics, and MM-PBSA Calculations, Frontiers in Chemistry. 8 (2020) 2296-2646. https://doi.org/10.3389/fchem.2020.584894

[17] H.J. Yang, M.J. Kim, S. Kang, N.R. Moon, D.S. Kim, N.R. Lee, K.S. Kim, S. Park, Topical treatments of Saussurea costus root and Thuja orientalis L. synergistically alleviate atopic dermatitis-like skin lesions by inhibiting protease-activated receptor-2 and NF-κB signaling in HaCaT cells and Nc/Nga mice, J Ethnopharmacol. 199:6 (2017) 97-105. doi: 10.1016/j.jep.2017.01.055.

[18] M.M. Pandey, S. Rastogi, A.K.S. Rawat, Saussurea costus: Botanical, chemical and pharmacological review of an ayurvedic medicinal plant. Journal of Ethnopharmacology. 110:3 (2007) 79–90. https://doi.org/10.1016/j.jep.2006.12.033.

[19] S. Ansari, Ethnobotany and pharmacognosy of qust/kut (Saussurea lappa, CB Clarke) with special reference of Unani medicine, Pharmacogn Rev. 13:26 (2019) 71-76. DOI:10.5530/phrev.2019.2.7

[20] N.M. Nasukhova, D.A. Konovalov, V.N. Orobinskaya, E.V. Galdin, Determination of biologically active compounds (costunolide and dehydrocostuslactone) in the leaves of some

forms of laurel noble (sweet bay), IOP Conf. Series: Earth and Environmental Science 941 (2021) 012015. https://doi.org/10.1088/1755-1315/941/1/012015.

[21] S.R. Ribone, S.A. Paz, C.F. Abrams, M.A. Villarreal, Target identification for repurposed drugs active against SARS-CoV-2 via high-throughput inverse docking. J Comput Aided Mol Des 36 (2022) 25–37. https://doi.org/10.1007/s10822-021-00432-3.

[22] H. Hajji, K. El Khatabi, H. Zaki, F. En-nahli, L. Hajji, T. Lakhlifi, M.A. Ajana, M. Bouachrine, Assessment of asthma treatment against SARS CoV-2 by using a computer approach, E3S Web Conf. 319 (2021) 01024. https://doi.org/10.1051/e3sconf/202131901024.

[23] C. Mchiri, H. Edziri, H. Hajji, M. Bouachrine, S. Acherar, C. Frochot, H.O. Badr Eldine, S.Ben Moussa, H. Nasri, 2-Aminopyridine Cadmium (II) meso-chlorophenylporphyrin coordination compound. Photophysical properties, X-ray molecular structure, antimicrobial activity, and molecular docking analysis, J Chem Sci. 134 (2022) 22. https://doi.org/10.1007/s12039-021-02022-0.

[24] H. Hajji, F. En-nahli, O. Abdessadak, K. El Khatabi, T. Lakhlifi, M.A. Ajana M. Bouachrine Antiproliferative Activity: Discovery of new Benzoxanthenes derivatives by Using Various Statistical Methods 2D/3D-QSAR and Molecular Docking, RHAZES: Green and Applied Chemistry, 12 (2021) 40-59. https://doi.org/10.48419/IMIST.PRSM/RHAZES-V12.26038.

[25] K. EL Khatabi, I. Aanouz, M. Alaqarbeh, M.A. Ajana, T. Lakhlifi, M. Bouachrine, Molecular docking, molecular dynamics simulation, and ADMET analysis of levamisole derivatiyes against the SARS-CoV-2 main protease (M Pro), Bioimpacts. 12 (2022) 107-113. https://doi.org/10.34172/bi.2021.22143.

[26] A. Belhassan, H. Zaki, S. Chtita, M. Alaqarbeh, N. Alsakhen, M. Benlyas, T. Lakhlifi, M. Bouachrine, Camphor, Artemisinin and Sumac Phytochemicals as inhibitors against COVID-19: Computational approach, Comput. Biol. Med. 136 (2021) 104758. https://doi.org/10.1016/j.compbiomed.2021.104758.

[27] J.B. Tong, X. Zhang, D. Luo, S. Bian, Molecular design, molecular docking and ADMET study of cyclic sulfonamide derivatives as SARS-CoV-2 inhibitors, Chinese J. Anal. Chem. 49:12 (2021) 63–73. https://doi.org/10.1016/j.cjac.2021.09.006.

[28] M.R.F. Pratama, H. Poerwono, S. Siswodiharjo, ADMET properties of novel 5-O-benzoylpinostrobin derivatives, JBCPP. 30:6 (2019) 20190251. https://doi.org/10.1515/jbcpp-2019-0251.

[29] M. Rbaa, A. Oubihi, H. Hajji, B. Tüzün, A. Hichar, E.H. Anouar, Elyor Berdimurodov, M.A. Ajana, A. Zarrouk, B. Lakhrissi, Synthesis, bioinformatics and biological evaluation of

novel pyridine based on 8-hydroxyquincline derivatives as antibacterial agents: DFT, molecular docking and ADME/T studies, J. Mol. Struct. 1244 (2021) 130934. https://doi.org/10.1016/j.molstruc.2021.130934.

[30] Í.F. Protti, D.R. Rodrigues, S.K. Fonseca, R.J. Alves, R.B. Oliveira, V.G. Maltarollo. Do Drug-likeness Rules Apply to Oral Prodrugs, ChemMedChem. 16:9 (2021) 1446–1456. https://doi.org/10.1002/cmdc.202000805.

[31] X. Chen, H. Li, L. Tian, Q. Li, J. Luo, Y. Zhang, Analysis of the Physicochemical Properties of Acaricides Based on Lipinski's Rule of Five. J. Comput. Biol. 27:9 (2020) 1397–1406. https://doi.org/10.1089/cmb.2019.0323.

[32] H. Hajji, K. Tabti, F. En-nahli, S. Bouamrane, T. Lakhlifi, M. A. Ajan, M. Bouachrine, In Silico Investigation on the Beneficial Effects of Medicinal Plants on Diabetes and Obesity: Molecular Docking, Molecular Dynamic Simulations, and ADMET Studies, Biointerface Res Appl Chem. 11:5 (2022) 6933–6949. https://doi.org/10.33263/BRIAC125.69336949.

[33] U. Norinder, C.A.S. Bergström, Prediction of ADMET Properties, ChemMedChem, 9:1 (2006) 920–937. https://doi.org/10.1002/cmdc.200600155.

[34] D.A. Abdelrheem, A.A. Rahman, K.N.M. Elsayed, H.R. Abd El-Mageed, H.S. Mohamed, S.A. Ahmed, Isolation, characterization, in vitro anticancer activity, DFT calculations, molecular docking, bioactivity score, drug-likeness and ADMET studies of eight phytoconstituents from brown alga sargassum platycarpum, J. Mol. Struct. 1225 (2021) 129245. https://doi.org/10.1016/j.molstruc.2020.129245.

[35] S. Zhang, K. Amahong, C. Zhang, F. Li, J. Gao, Y. Qiu, F. Zhu, RNA–RNA interactions between SARS-CoV-2 and host benefit viral development and evolution during COVID-19 infection, Brief Bioinform. 23(1) (2022) bbab397, https://doi.org/10.1093/bib/bbab397

[36] S. Zhang, K. Amahong, X. Sun, X. Lian, J. Liu, H. Sun, Y. Lou, F. Zhu, Y. Qiu, The miRNA: a small but powerful RNA for COVID-19, Brief Bioinform, 22(2) (2022) 1137-1149. DOI: 10.1093/bib/bbab062

[37] M.R.F. Pratama, H. Poerworo, S. Siswodihardjo, Molecular docking of novel 5-O-benzoylpinostrobin derivatives as wild type and L858R/T790M/V948R mutant EGFR inhibitor. JBCPP, 30(6) (2019) 20190301. https://doi.org/10.1515/jbcpp-2019-0301

[38] Y. Zhang, J.B. Ying, J.J. Hong, F.C. Li, T.T. Fu, F. Y. Yang. G.X. Zheng, X.J. Yao, Y. Lou, Y. Qiu, W.W. Xue, F. Zhu, How does chirality determine the selective inhibition of histone deacetylase 6? A lesson from trichostatin a enantiomers based on molecular dynamics.

ACS Chem. Neurosci, 10(5) (2019) 2467-2480. https://doi.org/10.1021/acschemneuro.8b00729

[39] W. Xue, F. Yang, P. Wang, G. Zheng, Y. Chen, X. Yao, F. Zhu, What Contributes to Serotonin–Norepinephrine Reuptake Inhibitors' Dual-Targeting Mechanism? The Key Role of Transmembrane Domain 6 in Human Serotonin and Norepinephrine Transporters Revealed by Molecular Dynamics Simulation, ACS Chem. Neurosci, 9 (5) 2018 1128-1140.

[40] W. Xue, P. Wang, G. Tu, F. Yang, G. Zheng, X. Li, X. Li, Y. X. Y. Chen, F. Zhu, Computational identification of the binding mechanism of a triple reuptake inhibitor amitifadine for the treatment of major depressive disorder. Phys Chem Chem Phys. 20(9) (2018) 6606-6616. doi: 10.1039/c7cp07869b.

Chapter 2

Assessment of asthma treatment against SARS CoV-2 by using a computer approach

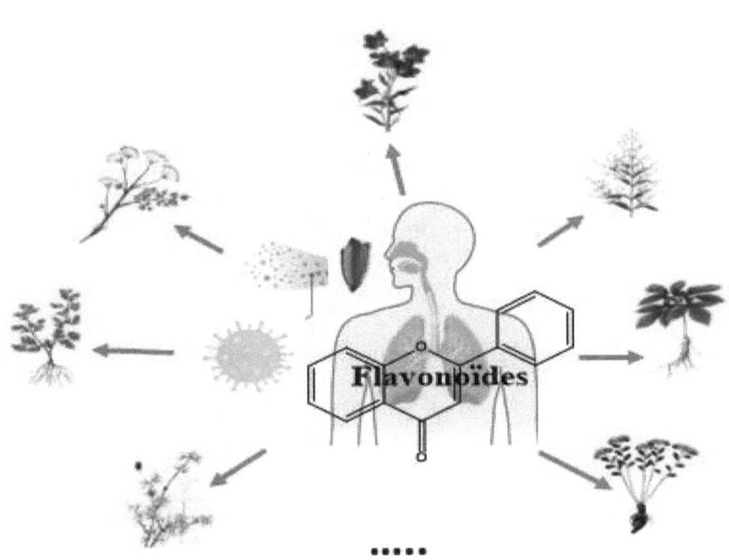

Abstract :

The disease caused by the coronavirus is called COVID-19, the degree of infection varies from person to another. According to the data collected to date, people with asthma and obesity are over-represented among adults hospitalized for COVID-19. The reason is very simple : COVID-19 is a disease that particularly attacks the respiratory system, including the lungs, This pandemic has led us to return to plants. Modern medicine has found its success thanks to traditional medicine, the effectiveness of which comes from medicinal plants. Currently, in China many people believe in the miraculous power of plants, boosting their immunity to protect against asthma. Therefore, this work aimed to study components of natural origin that have an anti-asthma effect that can be considered as the panacea against Covid-19, by using the most important method which is docking molecular . In this research, we performed a molecular docking study on molecules naturally occurring molecules based on the recently crystallized SARS CoV-2 protein (pdb code 7C6S). The results indicate that polyphenoles may be offered as a treatment against SARS CoV-2.

Keywords : Asthma, covid-19, Medicinal Plants, polyphenoles, Molecular Docking, ADMET.

1. Introduction

Today, Respiratory diseases are a big problem in society, best known to the general public of asthma and Covid-19 . Otherwise ,asthma is a chronic disease that affects humans due to inflammation of the airways in the lungs [1], When the muscle around the bronchi contracts and a lot of phlegm builds up in the airways,causing a blockage, it includes symptoms mainly manifested by a cough throughout the day and leads to wheezing, people with asthma may have trouble sleeping, fatigue and poor concentration, dyspnea, These symptoms are most severe at night and are variable over time in frequency and from person to person [2-3].

At the start of the SARS-COV-2 coronavirus pandemic [4] the medical profession was really afraid of seeing very many asthma patients infected with the virus because as we know COVID-19 is caused by a virus that directly affects the lungs so they have several symptoms in common, but quickly there are studies done in China that showed that there was not a significant proportion of asthma among people hospitalized because of COVID-19 [5-7]. On the other hand, there are exceptions for people with asthma if they have relatively severe asthma that is poorly controlled and is also a possible component of Chronic Obstructive Pulmonary Disease (COPD) in these cases they may be more at risk of having complications from a more severe COVID-19 disease [8].

In addition, humans have mainly used plants for treatment, thanks to the benefit of traditional Chinese medicine (TCM) which plays an effective role in the treatment of asthma may also give hope to control or help reduce inflammation in patients with COVID-19 [9]. However, in this research, we shed light on the effectiveness of polyphenols extracted from Chinese plants which have anti-inflammatory properties that could fight against COVID-19 infection [10-11], from computational studies which focus on the interactions between ligand-receptor by molecular docking method, as well as the toxicity study of these plants from ADMET [12].

2. Materials and methods

2-1. Selection of bioactives from polyphenols medicinal plants

In this research, we focused on 11 compounds based on polyphenols extracted from different Chinese medicinal plants, Figure 1 shows the structures of the molecules studied as well as the latter were considered to perform the molecular docking study [13].

Andrographolide (44)　　　**Carvacrol (45)**　　　**Chrysophanic Acid (46)**

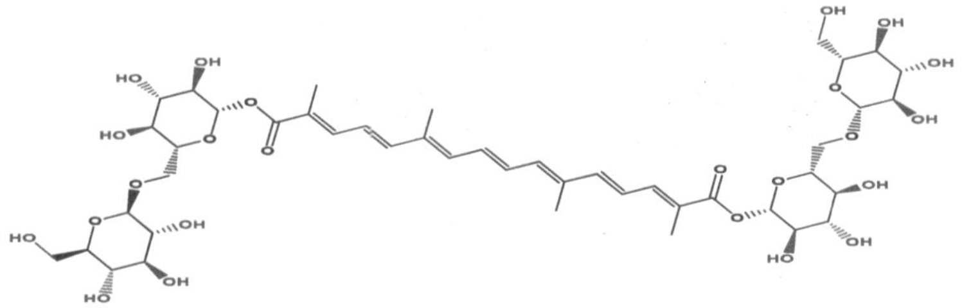

Crocin (47)

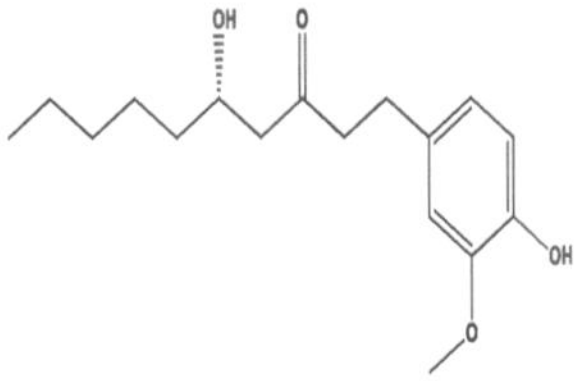

Gingerol (48)

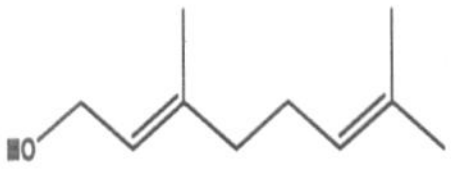

Geraniol (49)

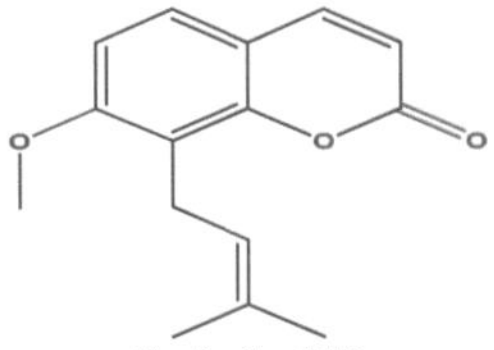

Osthole (50)

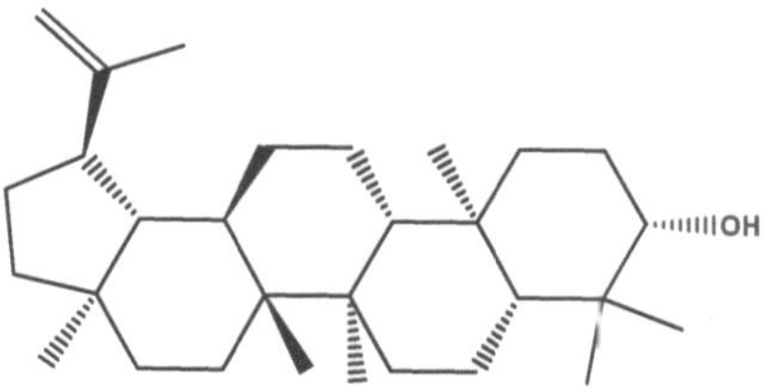

Lupeol (51)

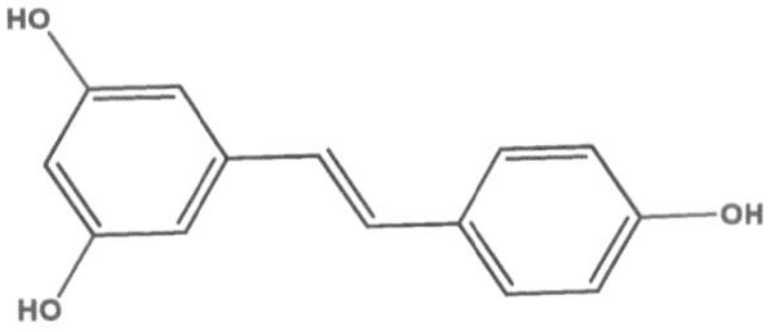

Resveratrol (52)

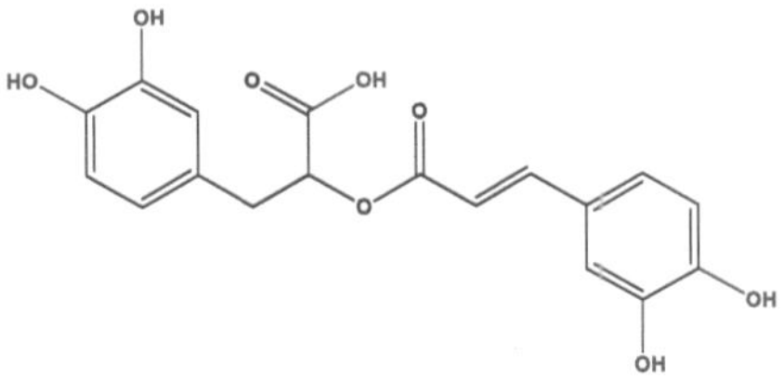

Rosmarinic acid (53)

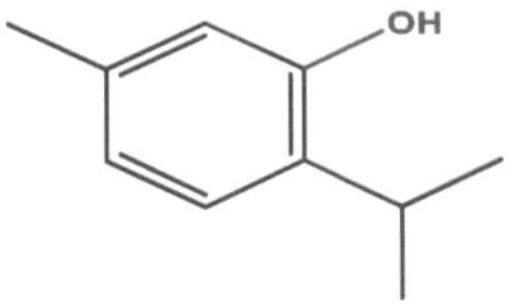

Thymol (54)

Fig 1 : Chemical Composition of several plants polyphenols (44-54)

2-2. Molecular docking :

Several software programs exist to carry out Molecular Docking, the main purpose of which is to predict the position of a ligand in its binding pocket. In addition in this research we downloaded the protein from protein data bank (https://www.rcsb.org/) and we removed the water molecules and converted the protein to PDB form by using discovery studio software. While its preparation with ligand is done by the SYBYL program, then we pooled the complex by using PYMOL software and we visualized the interactions by discovery studio [14-16].

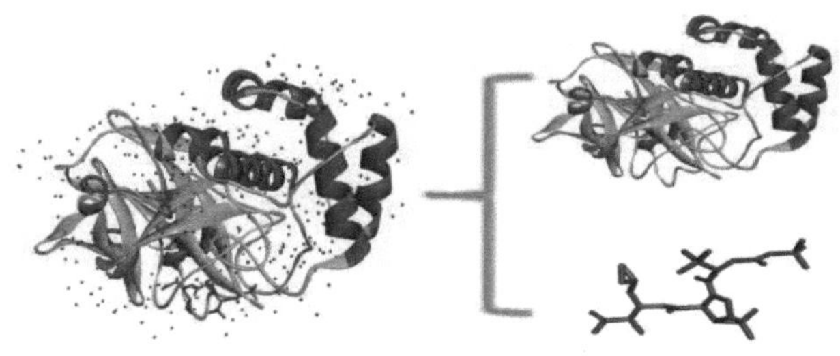

Fig. 2. Crystal structure of SARS CoV-2 main protease (PDB ID: 7C6S), Co-crystallized ligand and prepared protein for docking.

2-3. ADMET :

The study of plants to become drugs in the body over time is called pharmacokinetics. The method of using ADME type filters (Absorption, Distribution, Metabolism, Elimination) and Toxicology which is complementary to the ADME characteristics, opens up the possibility of filtering and selecting the molecules according to deferent parameters and their potentilla toxicity. As well as , Lipinski's rules allowing rapid identification on a large scale of molecules of a "drug-like" nature more likely to exhibit bioavailability characteristics [17-18].

3. Results and discussion

3-1. Molecular docking

We carried out molecular docking on 11 molecules of different medicinal plants by the Surflex-dock method by using Sybyl, to find the different types of interactions that are visualized by using discovery software, the results are grouped together in this table which show the strong interactions of the plants which can inhibit the SARS-CoV-2. In this docking analysis we applied the protein 7C6S which has a good Resolution: 1.60 Å on these Polyphenols, Moreover ,MET 49, ASP 187, TYR 54, GLY 138, GLU 166 and ARG 188 were found to be the main residues of hydrogen bond interactions for Crocin, it is the largest molecule compared to the others that we are studying, while the final structure of Lupeol was changed during molecular docking as it can be explained either by the molecule being broken or there is a chemical reaction that took place during this ligand-receptor docking. We noticed that Geraniol made two strong hydrogen bond interactions with THR 190 and ARG 188 residues with small distances 2.16 and 2.48 Å respectively, and interactions of alkyl and Pi-Alkyl with amino acids HIS 41, MET 49 and MET 165. Similarly we visualized the interactions on Carvacrol almost similar to the previous plant two hydrogen bond interactions and two interactions of Alkyl and Pi-Alkyl the stabilization of the latter two complexes often facilitates from these types of interactions. For Chrysophanic Acid we observed two interactions of Hydrogen Bond with THR

190 and ARG188 with distances ranging from 2.14 to 2.59 Å and only Pi-Alkyl interaction with residue ALA 191. Regarding Tymol and Osthole, this study revealed that there are other types of interactions apart from Hydrogen bond an interaction observed between Pi-Lone Pair with THR 190 from Tymol and Pi-Sigma which forms a bond with residue HIS 41 .For OstholeOn the other hand, we excluded in our research the components Gingerol, Resveratrol, Rosmarinic acid, Andrographolide because of the presence of unfavorable interactions on different regions of these plants.

Table 1. Different interactions and key residues (amino acids) for the inhibitor of SARS-CoV2.

Plant	Amine acid	Position	Distance	Interaction type	H-Bond	interaction
Crocin (47)	MET	49	2.18	Conventional Hydrogen bond		
	ASP	187	3.01			
	TYR	54	2.76			
	GLY	138	1.99			
	GLU	166	2.10			
	ARG	188	2.61			
	GLN	189	2.60	Carbon Hydrogen Bond		
	THR	169	2.59			
	GLY	170	2.48			
	HIS	163	4.08	Alkyl Pi-Alkyl		
	HIS	172	3.16			
	LEU	141	3.87			
	CYS	145	4.93			
Lupeol (51)	HIS	41	2.54	Conventional Hydrogen bond		
	HIS	164	3.0			
	CYS	145	2.94			
	GLY	143	2.78			
	GLU	166	2.92			
	GLN	189	3.62	Carbon Hydrogen Bond		
	MET	165	5.36	Alkyl Pi-Alkyl		
	MET	49	4.16			

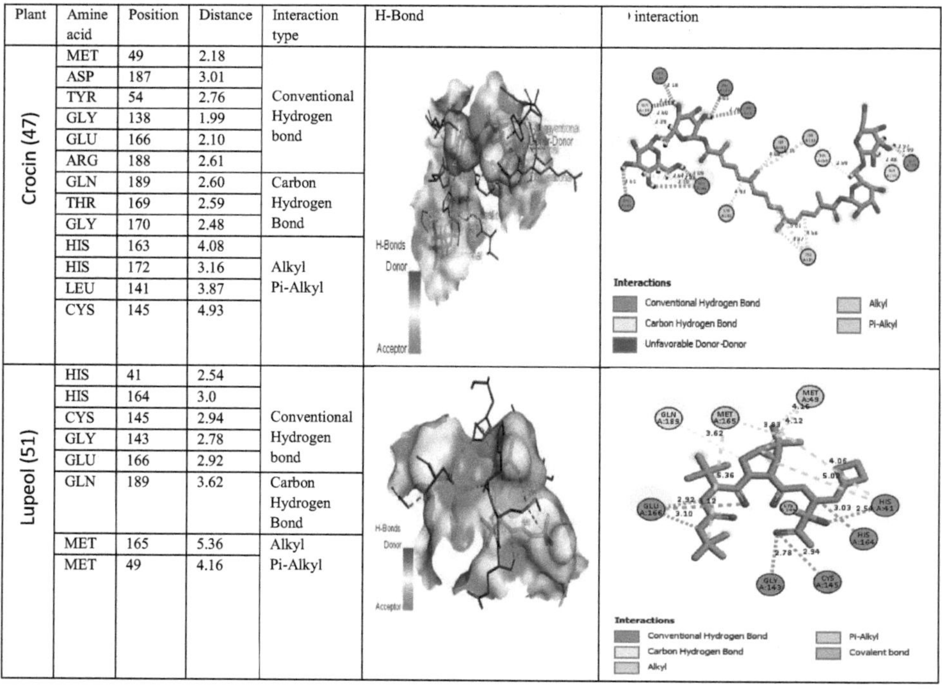

Compound	Residue	No.	Distance (Å)	Interaction Type		
Geraniol (49)	THR	190	2.16	Conventional Hydrogen Bond		
	ARG	188	2 .48			
	HIS	41	4.19			
	MET	49	3.53	Alkyl Pi-Alkyl		
	MET	165	4.06			
Carvacrol (45)	THR	190	2.84	Conventional Hydrogen Bond		
	ARG	188	1.96			
	MET	49	4.51			
	MET	165	5.26	Alkyl Pi-Alkyl		
Chrysophanic Acid (46)	THR	190	2.59	Conventional Hydrogen Bond		
	ARG	188	2.14			
	ALA	191	4.57	Pi-Alkyl		
Thymol (54)	GLN	192	3.00	Conventional Hydrogen Bond		
	THR	190	2.87	Pi-Lone Pair		
	MET	165	4.90	Alkyl		
	MET	49	4.16	Pi-Alkyl		
Osthole (50)	THR	190	2.85	Conventional Hydrogen Bond		
	GLN	189	2.75	Carbon Hydrogen Bond		
	HIS	41	2.85	Pi-Sigma		
	MET	49	4.16	Alkyl Pi-Alkyl		

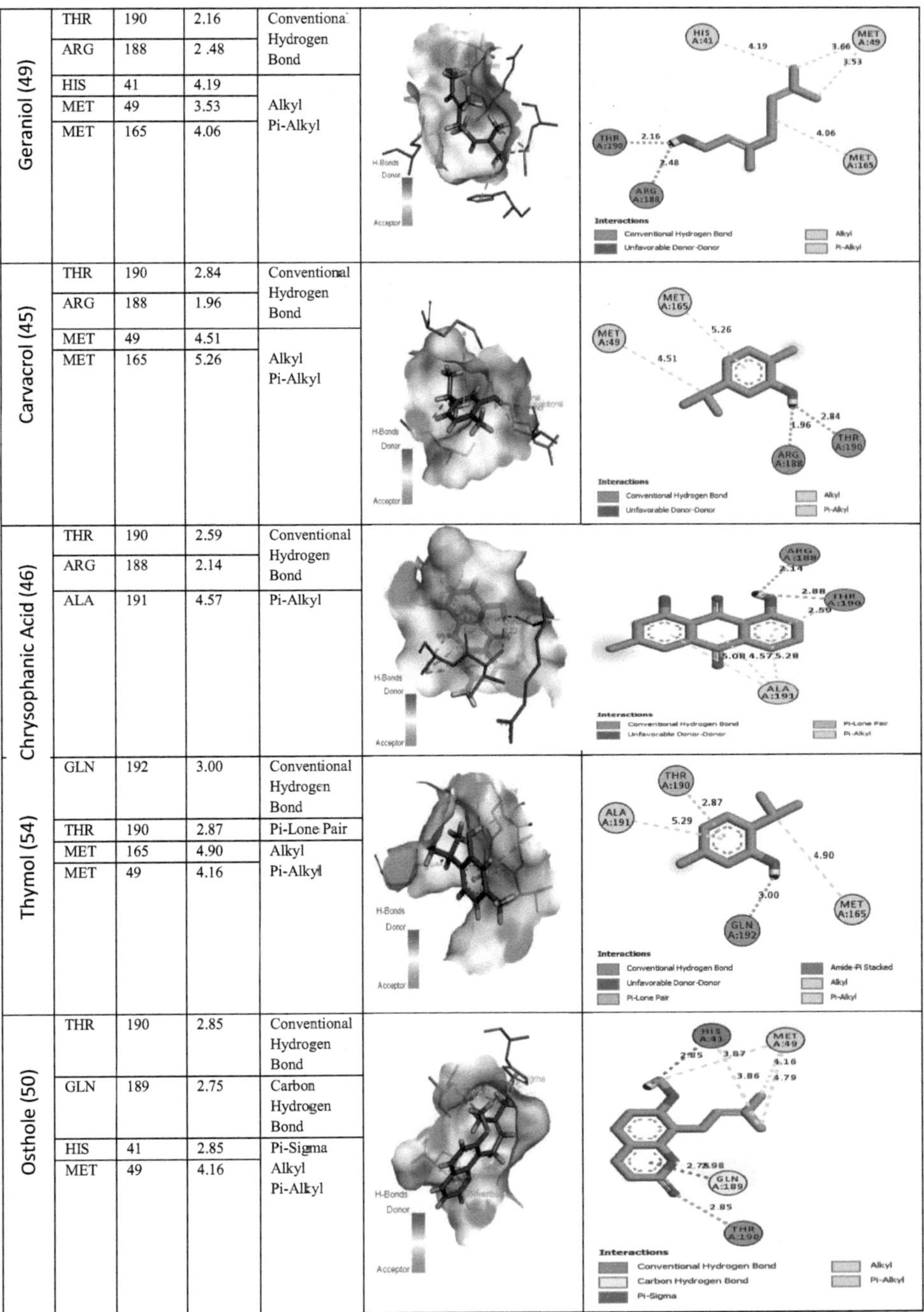

3-2. ADMET study

We performed the in silico ADMET study whose interest is to eliminate weak or toxic compounds and examine the ability of those that can become a drug from respecting ADMET parameters and Lipinski rules. The following table shows the different parameters of ADMET for molecules that gave good interactions in molecular docking such as Crocin, Lupeol, Geraniol, Carvacrol, Chrysophanic Acid, Thymol and Osthole. From the results we noticed that Crocin has an absorbance value of 0 which means that this molecule cannot be absorbed by the human intestine, as well as for logBB <-1 this indicates that it is poorly distributed to the brain and does not also comply with the five rules of Lipinski Tables 3. In the same way for Lupeol the parameter logP exceeds 5 for the rule of Lipinski. On the contrary, for the molecules Geraniol, Carvacrol, Chrysophanic Acid, Thymol and Osthole, these results presented in the Table 2 revealed that all these components had very good absorption, which is greater than 70%, and good Distribution and Exceretion and what concerns the results at the level of the AMES test of Toxicity they indicate that just the molecules of Chrysophanic Acid and Osthole are active as toxic but the others are not toxic and respect the five rules of Lipinski .

Table 2 : the results of the ADMET test with pKCSM of all compounds.

Molecules	Absorption	Distribution		Metabolism				Excretion	Toxicity	
	Intestinal absorption (human)	Blood Brain Barrier Permeability	CNS permeability	CYP				Renal OCT2 substrate	AMES toxicity	Hepatotoxicity
				2D6	3A4	2D6	3A4			
				Substrate		Inhibitor				
	Numeric (% Absorbed)	Numeric (log BB)	Numeric (log PS)	Categorical (Yes/No)				Categorial (Yes/No)	Categorical (Yes/No)	
Andrographolide (44)	94.442	-0.729	-2.62	No	Yes	No	No	No	Yes	No
Carvacrol (45)	93.712	0.381	-1.438	No	No	No	No	No	No	No
Chrysophanic Acid (46)	97.092	0.185	-2.089	No	No	No	No	No	Yes	No
Crocin (47)	0	-2.673	-6.992	No	No	No	No	Yes	No	No
Gingerol (48)	92.098	-0.506	-2.772	No	No	No	No	No	No	No
Geraniol (49)	93.467	0.621	-2.179	No	No	No	No	No	No	No
Osthole (50)	96.049	0.162	-2.183	No	Yes	No	Yes	No	Yes	No
Lupeol (51)	99.868	0.739	-1.35	No	Yes	No	No	No	No	No
Resveratrol (52)	89.057	-0.041	-2.098	No	No	Yes	Yes	No	No	No
Rosmarinic acid (53)	53.567	-1.547	-3.567	No	No	No	No	No	No	No
Thymol (54)	93.24	0.336	-1.349	No	No	No	No	No	No	No

Table 3 : Lipinski's role of the all compounds in the dataset

Compounds	Molecular weight (g/mol)	LogP	H-bond acceptors	H-bond donors	Rotatable bonds
Andrographolide (44)	350.455	1.96	5	3	3
Carvacrol (45)	150.221	2.824	1	1	1
Chrysophanic Acid (46)	254.241	2.181	4	2	0
Crocin (47)	976.972	-5.225	24	14	18
Gingerol (48)	294.391	3.233	4	2	10
Geraniol (49)	154.253	2 .671	1	1	4
Osthole (50)	244.29	3.310	3	0	3
Lupeol (51)	426.729	8.024	1	1	1
Resveratrol (52)	228.247	2	3	3	2
Rosmarinic acid (53)	360.318	1.761	7	5	6
Thymol (54)	150.221	2.824	1	1	1
Reference ligand	150 to 500	-0.7 to +5.0	≤ 10	≤ 5	≤ 9

Conclusion

In this work we looked at the hot topic asthma and covid-19, the link between these two diseases is strong it is the attack of the respiratory system, this is why we thought this time to turn to traditional Chinese medicine which pushed us to look for the polyphenols which are known in the literature for their effectiveness on asthma and to find the compounds that can be used in the management of SARS-Cov2 infections. For this reason, we carried out the molecular docking study with the main protease of SARS-CoV-2 on 11 polyphenol-based molecules, the results are satisfactory, this study helps us with the analyzes of ADMET to propose some plant such as Geraniol, Carvacrol, Chrysophanic Acid, Thymol and Osthole as a treatment for COVID-19 infections, these plants do not have the risk of toxicity so they respect Lipinski's rules. For further study,the synthesis of these compounds and the examination of their in vitro activity against SARS-CoV-2 could be of interest .

References

[1] BEURNIER, Antoine, JUTANT, Etienne-Marie, JEVNIKAR, Mitja, et al.Characteristics and outcomes of asthmatic patients with COVID-19 pneumonia who require hospitalisation. European Respiratory Journal, 2020, vol. 56, no 5. https://doi.org/10.1183/13993003.01875-2020

[2] MORAIS-ALMEIDA, Mário, AGUIAR, Rita, MARTIN, Bryan, et al. COVID-19, asthma, and biologic therapies: what we need to know. World Allergy Organization Journal, 2020, p. 100126. https://doi.org/10.1016/j.waojou.2020.100126

[3] HUGHES-VISENTIN, Alexzandra et PAUL, Anthea B. Mahesan. Asthma and COVID-19: What do we know now. Clinical Medicine Insights: Circulatory, Respiratory and Pulmonary Medicine, 2020, vol. 14, p. 1179548420966242. https://doi.org/10.1177/1179548420966242

[4] Zhu N., Zhang D., Wang W., Li X., Yang B., Song J., Zhao X., Huang B., Shi W., Lu R., Niu P., Zhan F., Ma X., Wang D., Xu W., Wu G., Gao G.F., Tan W. China novel coronavirus investigating and research team. A novel coronavirus from patients with pneumonia in China, 2019. N. Engl. J. Med. 2020;382(8):727-733. https://doi.org/10.1056/NEJMoa2001017

[5] World Allergy organization (WAO) https://www.worldallergy.org/COVID-19Information (2020), Accessed 27th Mar 2020

[6] D. Halpin, R. Faner, O. Sibila, J. Badia, A. Agusti Do chronic respiratory diseases or their treatment affect the risk of SARS-CoV-2 infection? Lancet Respir Med (2020), 10.1016/S2213-2600(20)30167-3 https://doi.org/10.1016/S2213-2600(20)30167-3

[7] X. Li, S. Xu, M. Yu, et al.Risk factors for severity and mortality in adult COVID-19 inpatients in Wuhan J Allergy Clin Immunol (2020 Apr 12), 10.1016/j.jaci.2020.04.006 pii: S0091-6749(20)30495-4

[8] MORAIS-ALMEIDA, Mário, PITÉ, Helena, AGUIAR, Rita, et al. Asthma and the coronavirus disease 2019 pandemic: a literature review. International archives of allergy and immunology, 2020, vol. 181, no 9, p. 680-688. https://doi.org/10.1159/000509057

[9] REN, Jun-ling, ZHANG, Ai-Hua, et WANG, Xi-Jun. Traditional Chinese medicine for COVID-19 treatment. Pharmacological research, 2020, vol. 155, p. 104743. https://doi.org/10.1016/j.phrs.2020.104743

[10] LEVY, Emile, DELVIN, Edgard, MARCIL, Valérie, et al. Can phytotherapy with polyphenols serve as a powerful approach for the prevention and therapy tool of novel coronavirus disease 2019 (COVID-19)?. American Journal of Physiology-Endocrinology and

Metabolism, 2020, vol. 319, no 4, p. E689-E708. https://doi.org/10.1152/ajpendo.00298.2020

[11] GIOVINAZZO, Giovanna, GERARDI, Carmela, UBERTI-FOPPA, Caterina, et al. Can Natural Polyphenols Help in Reducing Cytokine Storm in COVID-19 Patients?. Molecules, 2020, vol. 25, no 24, p. 5888. https://doi.org/10.3390/molecules25245888

[12] WU, Hao, WANG, J. Q., YANG, Y. W., et al. Preliminary exploration of the mechanism of Qingfei Paidu decoction against novel coronavirus pneumonia based on network pharmacology and molecular docking technology. Acta Pharmaceutica Sinica, 2020, vol. 55, no 3, p. 374-383.

[13] LIU, Jun-Xi, ZHANG, Yang, YUAN, Hong-Yu, et al. The treatment of asthma using the Chinese Materia Medica. Journal of Ethnopharmacology, 2020, p. 113558. https://doi.org/10.1016/j.jep.2020.113558

[14] HAJJI, Halima, et al. Antiproliferative Activity: Discovery of new Benzoxanthenes derivatives by Using Various Statistical Methods 2D/3D-QSAR and Molecular Docking. RHAZES: Green and Applied Chemistry, 2021, vol. 12, p. 40-59. https://doi.org/10.48419/IMIST.PRSM/rhazes-v12.26038

[15] CHTITA, Samir, BELHASSAN, Assia, AOUIDATE, Adnane, et al. Discovery of potent SARS-CoV-2 inhibitors from approved antiviral drugs via docking and virtual screening. Combinatorial Chemistry & High Throughput Screening, 2021, vol. 24, no 3, p. 441-454. https://doi.org/10.2174/1386207323999200730205447

[16] AANOUZ, I., BELHASSAN, A., EL-KHATABI, K., et al. Moroccan Medicinal plants as inhibitors against SARS-CoV-2 main protease: Computational investigations. Journal of Biomolecular Structure and Dynamics, 2020, p. 1-9. https://doi.org/10.1080/07391102.2020.1758790

[17] KHAMOULI, Saida, BELAIDI, Salah, OUASSAF, Mebarka, et al. Multi-combined 3D-QSAR, docking molecular and ADMET prediction of 5-azaindazole derivatives as LRRK2 tyrosine kinase inhibitors. Journal of Biomolecular Structure and Dynamics, 2020, p. 1-14. https://doi.org/10.1080/07391102.2020.1824815

[18] HADNI, Hanine et ELHALLAOUI, Menana. 3D-QSAR, docking and ADMET properties of aurone analogues as antimalarial agents. Heliyon, 2020, vol. 6, no 4, p. e03580. https://doi.org/10.1016/j.heliyon.2020.e03580

General Conclusion

The pursuit of effective treatments for respiratory diseases, particularly in the context of the COVID-19 pandemic, has highlighted the valuable interplay between traditional medicine and modern scientific techniques. This book has explored the potential of natural remedies, specifically focusing on the bioactive compounds in Saussurea costus and polyphenols from Chinese medicinal plants, in the treatment of respiratory ailments such as COVID-19 and asthma.

Our research has identified three key bioactive molecules from Saussurea costus—Costunolide, Cynaropicrin, and Dehydrocostus Lactone—as potential therapeutic agents against COVID-19. Through reverse docking studies, we have determined that Cynaropicrin exhibits promising inhibitory action against the main protease of SARS-CoV-2, supported by stable molecular dynamics simulations and favorable Lipinski's Rule outcomes. The ease of synthesis and potential efficacy of Cynaropicrin suggest it as a candidate for further in vitro studies to develop new antiviral drugs.

Additionally, our investigation into the link between asthma and COVID-19 has driven us to examine the role of polyphenols in traditional Chinese medicine. Molecular docking studies and ADMET analyses of polyphenol-based molecules, including Geraniol, Carvacrol, Chrysophanic Acid, Thymol, and Osthole, have demonstrated their potential in managing SARS-CoV-2 infections.

These compounds not only show effectiveness in binding to the viral protease but also adhere to safety profiles, making them suitable for further synthetic and in vitro evaluation.

In conclusion, the integration of natural remedies and computational chemistry offers a promising avenue for discovering new treatments for respiratory diseases. By combining the wisdom of traditional medicine with the precision of modern scientific methods, we can enhance our ability to combat complex health challenges. This interdisciplinary approach underscores the potential of harnessing nature's pharmacy alongside cutting-edge technology to improve public health outcomes.

More
Books!

yes
I want morebooks!

Buy your books fast and straightforward online - at one of world's fastest growing online book stores! Environmentally sound due to Print-on-Demand technologies.

Buy your books online at
www.morebooks.shop

Kaufen Sie Ihre Bücher schnell und unkompliziert online – auf einer der am schnellsten wachsenden Buchhandelsplattformen weltweit! Dank Print-On-Demand umwelt- und ressourcenschonend produzi ert.

Bücher schneller online kaufen
www.morebooks.shop

info@omniscriptum.com
www.omniscriptum.com
OMNIScriptum

Printed by Books on Demand GmbH, Norderstedt / Germany